FLORA OF TROPICAL EAST AFRICA

CARYOPHYLLACEAE

(including Illecebraceae)

W. B. TURRILL

Herbs (annual or perennial) or subshrubs, rarely shrubs. Leaves simple, opposite or rarely spiral, with or without stipules. Inflorescences cymose, often loosely dichasial but sometimes secund or compact, frequently many-flowered but by reduction few-flowered or flowers solitary. Flowers, with few exceptions, actinomorphic, hermaphrodite or unisexual (the species then dioecious, monoecious, or polygamous), parts in fives or rarely in fours, perianth hypogynous or perigynous, often with an internode between the calyx and the corolla. Calyx of free sepals (polysepalous) or gamosepalous, frequently persistent and becoming more or less scarious. Corolla of free petals, in the gamosepalous genera with usually well differentiated lamina and claw and corona of scales often present, in the polysepalous genera not so differentiated, entire to deeply bi-lobed, sometimes absent. Stamens 5 + 5 or fewer by reduction. Gynoecium of 2, 3, 4, or 5 carpels, syncarpous ; ovary 2-, 3-, 4-, or 5-locular or unilocular, with numerous, few, or solitary ovules with axile, central, free central, or basal placentation. Fruits capsular or reduced to one-seeded indehiscent nutlets. Seeds with perisperm but no, or at most very little, endosperm ; embryo most often curved.

The petals, when present, are sometimes referred to as staminodes, especially in the first seven genera in the key below. There is experimental as well as morphological evidence of the close relationship (structural, onto-genetic, and genetic) between androecium (stamens) and corolla (petals) in the family, but for uniformity of treatment in this Flora it is considered advisable to use the terms corolla and petals for the whorl (when present) between sepals and functional stamens.

Sepals free one from another, hypogynous or on rim of a
   perigynous tube :
  Leaves with stipules :
    Fruits dehiscent (capsular) :
      Styles united, at least in lower part :
        Petals entire or slightly emarginate or dentate :
          Carpels (style arms and capsule valves) 5 .   1. **Krauseola**
          Carpels (style arms and capsule valves) 3 :
            Sepals keeled . . . . .   2. **Polycarpon**
            Sepals not keeled . . . . .   3. **Polycarpaea**
          Petals deeply bisegmented . . . .   4. **Drymaria**
        Styles free . . . .   5. **Spergula**
      Fruits indehiscent (one-seeded) :
        Subshrubs ; petals minute ; persistent filiform
          style . . . .   6. **Pollichia**
        Herbs ; petals equalling or slightly exceeding
          sepals ; stigma subsessile . . . .   7. **Corrigiola**

FIG. 1. *KRAUSEOLA GILLETTII*, from *Gillett* 13676—**1**, part of plant, × ½ ; **2**, flowering branch, × 1 ; **3**, node showing arrangement of leaves and stipules, × 6 ; **4**, flower, with some sepals bent back to show inside, × 8 ; **5 & 6**, sepals, × 8 ; **7**, petal, × 8 ; **8**, part of flower from inside to show arrangement of petals and stamens, × 16 ; **9**, gynoecium, × 16 ; **10**, capsule surrounded by persistent sepals, × 6 ; **11**, seed, × 24.

Leaves exstipulate :
  Capsules with the same number of teeth or valves
    as styles :
    Carpels (styles, capsular teeth or valves) 4 to 6 .   8. **Sagina**
    Carpels (styles, capsular teeth or valves) 3 . . 9. **Minuartia**
  Capsule with double the number of teeth or valves
    as styles :
    Carpels (styles) usually 4–5 ; petals emarginate
    or bifid .   .   .   .   . 10. **Cerastium**
    Carpels (styles) usually 2–3 :
      Petals entire or at most emarginate or dentate 11. **Arenaria**
      Petals deeply bilobed or bisegmented .   12. **Stellaria**
Sepals united to form a gamosepalous calyx :
  Calyx with 8–10 main veins or ribs (5 dorsal to the
    teeth, 5 commissural) :
    Capsule unilocular :
      Carpels alternating with the calyx teeth . . 13. **Uebelinia**
      Carpels opposite the calyx teeth .   .   14. **Melandrium**
    Capsule with septa, at least in lower part .   15. **Silene**
  Calyx with numerous parallel often obscure veins ;
    carpels (styles) 2 ; ovary unilocular .   .   16. **Dianthus**

1. KRAUSEOLA

Pax & Hoffm. in E. & P. Pf., ed. 2, 16c : 295, 308 (1934)
Pleiosepalum Moss in J. B. 69 : 65 (1931), non
Hand.-Mazz. (1922), *nom. illegit.*

Annual or perennial herbs. Stems approximately erect, branched. Leaves opposite (often with the false appearance of being in whorls), obovate to narrowly oblanceolate, entire, with scarious stipules. Flowers several to numerous, in rather compact to loose cymes. Sepals 9–12, spirally arranged, imbricate. Petals 5–8, hyaline-membranaceous, much shorter than the sepals. Fertile stamens 5–8, alternating with the petals. Ovary with numerous ovules ; style simple below with 5 stigmatic arms. Capsule 5-valved, with numerous seeds.

K. gillettii *Turrill* in K.B. 1954 : 533 (1955). Type : Kenya, Northern Frontier Province, Sololo Police Post, 2 Aug. 1952, *Gillett* 13676 (K, holo. !)

An annual or perennial herb, 9–27 cm. tall, branched, with the stems and leaves more or less hirsute with branched hairs interspersed with shorter ones. Leaves oblanceolate or narrowly oblanceolate, acute or subobtuse and shortly apiculate at apex, gradually narrowed at the base but not or scarcely petiolate, 8–47 mm. long, 2–13 mm. broad ; stipules elongate-triangular, gradually acuminate, 2–3 mm. long. Cymes 3–9-flowered, with the rhachis and pedicels shortly hirsute ; pedicels 2–5 mm. long. Sepals in outline lanceolate or narrowly ovate or lanceolate-elliptic, acutely apiculate, 3·5–5 mm. long, 1·25–2·5 mm. broad, the outer ones the smaller, margins broadly scariose, 3–7 nerved. Petals oblong, rounded or obscurely dentate at the apex, 1 mm. long. Filaments 1·5–1·75 mm. long ; anthers scarcely 1 mm. long. Ovary broadly ovoid ; style 1·5 mm. in total length including stigmatic arms. Capsule broadly ovoid, 6 mm. tall, 5 mm. in diameter. Seeds reniform, 0·75 mm. wide, shining. Fig. 1.

KENYA. Northern Frontier Province : Dandu, 1 May 1952, *Gillett* 12995 !
DISTR. **K1** ; endemic to northern Kenya
HAB. In *Acacia–Commiphora* scrub, 780–810 m.

E.M.S.

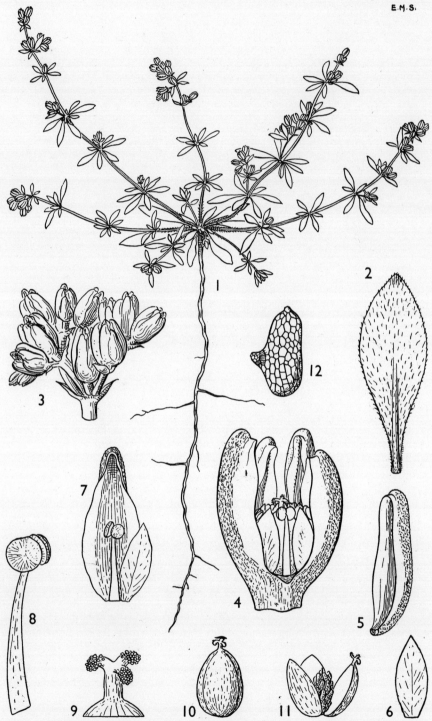

FIG. 2. *POLYCARPON PROSTRATUM*, from *Wild* 3054—**1**, plant, × 1 ; **2**, leaf, lower surface, × 4 ; **3**, inflorescence, × 4 ; **4**, flower with one sepal removed, × 14 ; **5**, sepal, × 14 ; **6**, petal, × 14 ; **7**, sepal, petal and stamen, × 14 ; **8**, stamen, × 30 ; **9**, style and stigmatic branches, × 30 ; **10**, capsule, × 14 ; **11**, dehisced capsule, showing seeds, × 14 ; **12**, seed, × 80.

2. POLYCARPON

L., Syst., ed. 10, 881 (1759)

Annual or perennial herbs. Leaves often apparently verticillate and with membranous stipules. Flowers small, in cymes terminating the numerous branches. Sepals keeled. Petals up to 5, entire or emarginate, shorter than the sepals. Stamens 3–5. Ovary unilocular, with numerous ovules ; style with three stigmatic branches. Capsule 3-valved.

P. prostratum (*Forsk.*) *Aschers. & Schweinf.* in Oesterr. Bot. Zeitschr. 39 : 128 (1889) ; Consp. Fl. Angol. 1 : 110 (1937) ; Milne-Redhead in K.B. 1948 : 451 (1949) ; Fl. Pl. Anglo-Egypt. Sudan 1 : 89 (1950) ; F.W.T.A., ed. 2, 1 : 131 (1954). Type : Egypt, near Cairo, *Forskåol* (C, holo.)

Apparently an annual plant with much branched more or less ascending stems ; lower internodes often pubescent, the upper with a line of hairs, or all internodes glabrescent to glabrous. Leaves opposite or verticillate, linear-oblanceolate, 0·5–2·5 cm. long, pubescent to glabrous, with ovate or lanceolate stipules. Inflorescences of many flowered cymes. Flowers sessile or lateral ones with pedicels up to 3 to 6 mm. long. Sepals 5, lanceolate, 3–4 mm. long, with a broad green median zone and transparent membranous margins. Petals sometimes 2–3, sometimes apparently absent, 1–1·5 mm. long, thin and translucent. Stamens (in flowers dissected) 3. Fig. 2.

TANGANYIKA. Shinyanga District : Shinyanga, *Koritschoner* 2196 !
DISTR. T1, 4 ; widely ranging in many tropical and some subtropical areas of both East and West Hemispheres, especially the latter
HAB. Damp or marshy places, sandy river banks

SYN. *Alsine prostrata* Forsk., Fl. Aegypt-Arab. 207 (1775)
   Pharnaceum depressum L., Mant. 2 : 562 (1771). Type : E. India (LINN, holo. !)
   Polycarpon depressum (L.) Rohrb. in Mart., Fl. Bras. 14 (2) : 257, t. 59 (1872);
      F.C.B. 2 : 148 (1951), *non* Nutt. (1838), *nom. illegit.*
   P. *loeflingii* (Wight & Arn.) Benth. in G.P. 1 : 153 (1862) ; F.T.A. 1 : 144
      (1868) ; F.W.T.A. 1 : 111 (1927), *nom. illegit.*

3. POLYCARPAEA

Lam. in Journ. Hist. Nat. Par. 2 : 3, t. 25 (1792)

Annual or perennial plants, mostly herbaceous but sometimes woody near the base. Stems erect or ascending, often much branched. Leaves linear or oblanceolate to ovate, opposite or apparently whorled, with scarious stipules. Inflorescences in terminal, loose or compacted, cymes. Flowers 5-merous ; perianth and stamens hypogynous or somewhat perigynous. Sepals entirely scarious, not keeled, white or brown to purple. Petals shorter than the sepals. Stamens 5 (or reduced from 5). Ovary unilocular with few to many ovules ; style with three short lobes or capitate. Capsule 3-valved.

Ovules usually 3 (2–4) :
 Inflorescences effuse, usually numerous ; sepals 1·5–
  2·25 mm. long. . . . . . . 1. *P. eriantha*
 Inflorescences spherically compact, usually solitary at
  the ends of branches ; sepals 2·5–3 mm. long . 2. *P. linearifolia*
Ovules usually 7–12 (5–13) :
 Leaves linear or narrowly linear :
  Style about 1·5 mm. long . . . . . 3. *P. tenuistyla*
  Style about 0·25 mm. long . . . . . 4. *P. corymbosa*
 Leaves narrowly oblanceolate . . . . . 5. *P. grahamii*

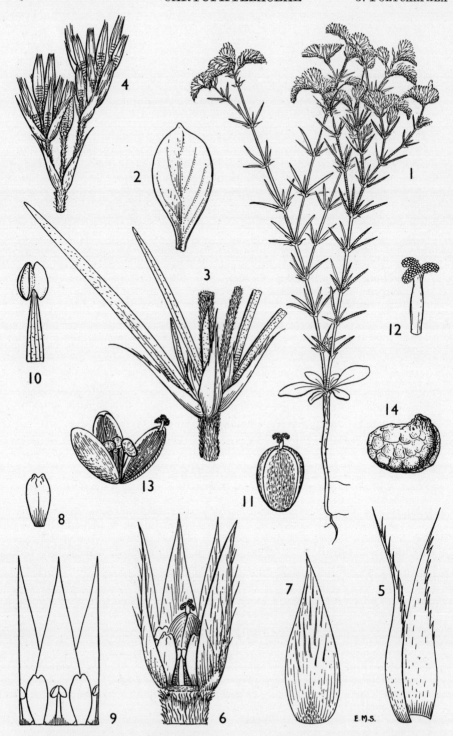

FIG. 3.   *POLYCARPAEA ERIANTHA*—**1**, plant, × ⅔ ; **2**, basal leaf, upper surface, × 3 ; **3**, node, showing leaves and stipules, × 8 ; **4**, part of inflorescence, × 5 ; **5**, bract, × 20 ; **6**, flower with one sepal removed, × 20 ; **7**, sepal, × 20 ; **8**, petal, × 20 ; **9**, part of flower from inside, showing arrangement of petals and stamens (diagrammatic), × 20 ; **10**, stamen, × 60 ; **11**, gynoecium, × 20 ; **12**, style and stigmatic lobes, × 120 ; **13**, dehisced capsule, × 20 ; **14**, seed, × 60. 1–13, from *A. S. Thomas 3545* ; **14**, from *Eggeling 2342*.

1. **P. eriantha** [*Hochst. ex*] *A. Rich.*, Tent. Fl. Abyss. 1 : 303 (1847) ; F.W.T.A. 1 : 112 (1927) & ed. 2 : 1, 132 (1954) ; Consp. Fl. Angol. 1 : 111 (1937) ; Fl. Pl. Anglo-Egypt. Sudan 1 : 87 (1950) ; F.C.B. 2 : 144 (1951). Type : Ethiopia, near Gapdia, *Schimper* 823 (K, iso. !)

An annual herb, branched from the base, 3–19 cm. tall, internodes covered with a woolly indumentum of more or less curled hairs. Leaves opposite or whorled ; basal leaves (most often absent in the flowering specimens examined) spathulate, 1–1·5 cm. long, 3–4 mm. broad ; cauline leaves linear, acute and then terminating in a hair-like appendage 1 mm. long and more or less caducous, 5–18 mm. long, 0·5–1 mm. broad, 1-nerved, with a woolly indumentum when young, becoming glabrous with age. Inflorescences terminal, of many-flowered cymes somewhat varying in density but single pedicelled flowers present between the main branches. Flowers silvery white to pink or reddish. Sepals lanceolate, acuminate, 2–2·25 mm. long, with white more or less curled hairs on the abaxial surface or glabrous. Petals about half the length of the sepals, slightly emarginate. Stamens very small, apparently often reduced to 2 or 3. Ovary with 3 ovules ; style 0·25 mm. long or even less. Fig. 3.

var. **eriantha**

Sepals hairy

UGANDA. Karamoja District : Moruatukan [? Moruaturukan], 24 May 1940, *A. S. Thomas* 3545 !
KENYA. S. Kavirondo District : Lambwe area, July 1934, *Napier* 6647 !
TANGANYIKA. Morogoro, May 1930, *Haarer* 1914 !
DISTR. U1–4 ; K1 (?), 5 ; T1, 4–8 ; most parts of tropical Africa
HAB. Weed of cultivated ground, roadsides, sandy wooded grassland, bushland, dry hillsides, 480–1650 m.

SYN. *P. corymbosa* (L.) Lam. var. *effusa* Oliv. in F.T.A. 1 : 145 (1868), *pro parte*

var. **effusa** (*Oliv. emend. Pax*) *Turrill* in K.B. 1954 : 503

Sepals glabrous

TANGANYIKA. Mwanza and Kwimba, June 1932, *Rounce* 54, in K ! (*Rounce* 54, in EA is *P. corymbosa*)
DISTR. T1, 3, 8 ; here and there within the range of var. *eriantha*
HAB. As for var. *eriantha*

SYN. *P. corymbosa* var. *effusa* Oliv. in F.T.A. 1 : 145 (1868) *pro parte*, emend. Pax in E.B.J. **17** : 590 (1893). Type : Nigeria, Nupe, *Barter* 808 (K, lecto. !)

Very small-flowered plants, with sepals 1·5 mm. long, were collected in Tanganyika, Moa, coconut plantation near wells, 4 Aug. 1953, *Drummond & Hemsley* 3644. They have been tentatively retained in *P. eriantha* var. *effusa* (see Hook. Ic. Pl. 36 : t. 3526 (1956)).

2. **P. linearifolia** (*DC.*) *DC.*, Prodr. 3 : 374 (1828) ; F.T.A. 1 : 146 (1868) ; F.W.T.A. 1 : 112, fig. 40 (1927) ; ed. 2, 1 : 133, fig. 47 (1954). Type : probably Senegal, Herb. Jussieu (P, holo.)

An annual herb, usually erect and rather strict, more or less branched with the branches often somewhat fastigiate, 2·8–4 dm. tall, internodes covered with a woolly indumentum of more or less curled hairs. Leaves opposite or falsely whorled ; cauline leaves linear, acute or subacute and then terminating in a hair- or bristle-like appendage up to 1 mm. long and more or less caducous, 5–16 mm. long, 0·5–1 mm. broad, 1-nerved, glabrous or with a few hairs, usually marginal, when young. Inflorescences strictly terminal to the branches, very compact, spherical to obloid. Flowers silvery white. Sepals lanceolate, acuminate, 2·5–3 mm. long, glabrous. Petals

1 mm. long, rounded at the apex. Stamens very small, typically 5. Ovary with 3 (rarely 4) ovules ; style 0·25 mm. long or even less.

TANGANYIKA. Tabora District : Urambo, 31 May 1950, *Moors* 30 ! ; Tunduru District : *Allnutt* 42 !
DISTR. **T4**, 8 ; probably widely spread in tropical Africa, but particularly common in the western parts
HAB. In bushland, sandy river beds, and weed of grasslands and waste places, 250–780 m.

SYN. *Paronychia linearifolia* DC. in Lam., Encycl. 5 : 26 (1804)
     Illecebrum linearifolium (DC.) Pers., Syn. 1 : 261 (1805)

3. P. tenuistyla *Turrill* in K.B. 1954 : 504 ; Hook. Ic. Pl. 36 : t. 3528 (1956). Type : Kenya, Kilifi District, " Ribe to Galla country," *Wakefield* (K, holo. !)

Probably perennial, at least with the lower parts of the stems more or less woody ; stems up to 2·2 dm. long, densely leafy, with hirsute internodes. Cauline leaves opposite or whorled, aggregated in axillary branches, narrowly linear, acute, mucronate, 3–11 mm. long, 0·25 mm. broad, spreading, glabrous or nearly glabrous. Inflorescences terminal, many flowered, more or less congested, 2–4·7 cm. broad. Flowers apparently white. Sepals lanceolate, acute or shortly acuminate, 3 mm. long, glabrous. Petals narrowly lanceolate, acute or irregularly 2–3-dentate, 2·3 mm. long. Stamens 5, 2 mm. long. Ovary with 10–12 ovules ; style slender, about 1·5 mm. long.

KENYA. Kilifi District : " Ribe to Galla country " (received) May 1880, *Wakefield* !
DISTR. **K7** ; endemic to south-eastern Kenya
HAB. Not known (probably bushland)

4. P. corymbosa (*L.*) *Lam.*, Ill. Gen. Encycl. 2 : 129 (1797) ; F.T.A. 1 : 145 (1868), *pro parte* ; F.W.T.A. 1 : 113 (1927) & ed. 2, 1 : 132 (1954) ; Consp. Fl. Angol. 1 : 110 (1937) ; Fl. Pl. Anglo-Egypt. Sudan 1 : 87 (1950) ; F.C.B. 2 : 145 (1951). Type : Ceylon, Herb. Hermann (BM, lecto. !)

An annual herb, generally erect and often very strict, often branched from the base but sometimes with simple main stems, 5–38 cm. tall, internodes covered with more or less curled white hairs when young, often glabrescent when older. Leaves opposite or apparently whorled, narrowly linear, acute and then terminating in a hair-like bristle 1 mm. long and caducous, 5–30 mm. long, 0·5–1 mm. broad, 1-nerved, glabrous or nearly so when mature. Inflorescences terminal to branches, of many-flowered cymes, differing greatly in density (see below). Flowers silvery white to pink or purplish red. Sepals lanceolate, acuminate, 2·5–3·75 mm. long, glabrous. Petals about 1·25 mm. long, slightly emarginate or erose. Stamens usually 5, 0·75 mm. long. Ovary with 5–13 ovules ; style 0·25 mm. long or even less.

UGANDA. West Nile District : Terego, March 1938, *Hazel* 461 !
KENYA. Machakos District : Makueni, 17 October 1947, *Bogdan* 1378 !
TANGANYIKA. Kondoa District : Babati–Kondoa road, 3 July 1950, *Bally* 7854 ! ; Mpanda District : Tumba, 13 March 1951, *Bullock* 3768 ! ; Shinyanga District : Shinyanga, *Koritschoner* 2107 !
DISTR. **U1**, 3, 4 ; **K1**, 4, 7 ; **T1**, 3, 4–6, 8 ; widely distributed in the tropics of both the East and West Hemispheres
HAB. Weed of cultivated ground, roadsides, grassland, and woodland, particularly on sandy soils, 900–1800 m.

SYN. *Achyranthes corymbosa* L., Sp. Pl. 205 (1753)

This species, as here accepted, is extremely variable, especially in the size, degree of robustness, and amount of branching of the plants and in the size, number of flowers, and compactness of the inflorescence. The extremes of inflorescence compactness are very striking and probably there are some gene differences involved. On the other hand, there are plants, from tropical Africa, showing every grade between the extremes.

On the whole, the most compact and often spherical inflorescences, with no or almost no inflorescence branches visible, are West African. The majority of the specimens from Tropical East Africa have more or less loose inflorescences or, at least, such as are not spherically compact as are the inflorescences of the extreme variants often accepted as *P. linearifolia* (DC.) DC. Prod. 3 : 374 (1828). A few specimens appear to link *P. corymbosa* with *P. linearifolia*, if the two be kept as different species. Examples of such intermediates are :

UGANDA. Teso District : Serere, Nov.–Dec. 1931, *Chandler* 92 ! ; West Nile District, Terego, Mar. 1938, *Hazel* 462 ! ; Imatong Mts., Laboni, 23 Dec. 1925, *A. S. Thomas* 1748 !
TANGANYIKA. Shinyanga District : near Shinyanga, 9 Mar. 1933, *Bax* 21 ! ; Rufiji District : Mafia Island, Adani, 9 Aug. 1937, *Greenway* 5039 !

S. Balle (F.C.B. 2 : 146–8, 1951), for the Belgian Congo, names and describes three varieties and three forms within *P. corymbosa*, apparently excluding *P. lineari(i)folia* (DC.) DC. She acknowledges that these are linked by intermediates and it would appear of doubtful value to give names to slight variants until modern methods of synthetic taxonomy, including cultural and genetical experiments, have been applied to the whole group.

5. **P. grahamii** *Turrill* in K.B. 1954 : 503 ; Hook. Ic. Pl. 36 : t. 3527 (1956). Type : Kenya, Kilifi District, Fundi Isa, *Graham* 1617 (K, holo. !, EA, iso. !)

An annual herb branched from the base, 20–30 cm. tall, internodes covered with a hirsute indumentum of more or less curled hairs. Leaves opposite ; basal leaves absent in the flowering specimens examined ; cauline leaves narrowly oblanceolate, rounded at the apex, gradually narrowed at the base, 4–10 mm. long, 1·5–2·5 mm. broad, nerves quite inconspicuous, hispid on the abaxial and somewhat hirsute to glabrescent on the adaxial surface ; stipules narrowly lanceolate, acuminate, scarious, 2–3 mm. long. Inflorescences terminal, of many flowered cymes which are more or less congested with usually single pedicelled flowers between the main branches. Flowers silvery white. Sepals lanceolate, acute or shortly acuminate, 3·5 mm. long, glabrous. Petals 2 mm. long, at the apex irregularly dentate or emarginate. Stamens 5, 1·75 mm. long. Ovary with 18–20 ovules ; style 0·75 mm. long.

KENYA. Kilifi District : Fundi Isa, 1929, *Graham* 1617 !
DISTR. **K**7 ; endemic to eastern Kenya
HAB. Sandy places in bushland near the shore

4. DRYMARIA

[Willd. ex] Roem. & Schultes, Syst. Veg. 5 : p. xxxi (1819)

Herbs. Leaves flat, opposite, with small stipules. Flowers small, solitary, in the axils of leaves or forming a cyme terminal to the branches ; receptacle slightly perigynous. Sepals (4–) 5, free. Petals (4–) 5, deeply bisegmented. Stamens 5 or fewer, filaments shortly connate at base. Ovary unilocular with 2–numerous ovules ; styles 3, connate below. Capsule with 3 valves.

D. cordata (*L.*) [*Willd. ex*] *Roem. & Schultes*, Syst. Veg. 5 : 406 (1819) ; F.T.A. 1 : 143 (1868) ; P.O.A. C : 177 (1895) ; F.W.T.A. 1 : 111, fig. 39 (1927) & ed. 2, 1 : 131, fig. 46 (1954) ; Consp. Fl. Angol. 1 : 109 (1937) ; F.P.N.A. 1 : 156 (1948) ; Fl. Pl. Anglo-Egypt. Sudan 1 : 86 (1950) ; F.C.B. 2 : 139 (1951)

A straggling herb with procumbent and more or less ascending branched stems, often rooting at the lower nodes, quadrangular, glabrous or papillose especially in the upper internodes ; internodes slender, generally 2–6 cm. long. Leaves opposite, blades ovate to very broadly ovate, acute or subacute and shortly apiculate or even rounded at apex, cordate, truncate, or shortly cuneate at base, 1 cm. long and 0·6 cm. broad up to 3·5 cm. long and 3·0 cm. broad, 3–7-palminerved, glabrous ; petioles 2–7 mm. long ; stipules inter-

FIG. 4. *DRYMARIA CORDATA*, from *Purseglove* 448—**1**, part of plant, × 1 ; **2**, node, showing leaf and stipular segments, × 2 ; **3**, part of inflorescence, × 4 ; **4**, sepal, × 4 ; **5**, flower with sepals removed, × 6 ; **6**, petal, × 6 ; **7**, stamens and gynoecium, × 6 ; **8**, gynoecium, × 6 ; **9**, fruit with persistent sepals, × 4 ; **10**, seed, × 6.

petiolar, about 1 mm. long, deeply fringed or composed of a number of subulate segments. Cymes terminal or axillary, often loose on long slender peduncles and with slender branches ; pedicels at anthesis mostly 3–5 mm. long. Sepals narrowly lanceolate, acute, 3·5–5 mm. long, keeled along the papillose midrib, with one nerve on each side ; the " flowers " (presumably the sepals) are said to be " sticky." Petals 3 mm. long, white. Ovules usually about 3. Seeds 1·5 mm. in greatest diameter, bluntly tubercled. Fig. 4.

UGANDA. Kigezi District : Kachwekano Farm, May 1949, *Purseglove* 2831 !
KENYA. Northern Frontier Province : Marsabit, 14 Feb. 1953, *Gillett* 15111 ! ; Embu District : Thiba River, 4 Nov. 1939, *Mrs. H. Copley in Bally* 424 !
TANGANYIKA. Lushoto District : Amani, 13 May 1950, *Verdcourt* 192 ! ; Morogoro District : Uluguru Mts., Lukwangule plateau, above Chenzema Mission, 13 Mar. 1953, *Drummond & Hemsley* 1562 !
DISTR. U1–4 ; K1, 3–5, 7 ; T2, 3, 5–7 ; pan-tropical
HAB. Grasslands, forest and bushland margins, roadsides and cultivated areas, often in shade or moist soil, 870–2700 m.

SYN. *Holosteum cordatum* L., Sp. Pl. 88 (1753). Type : " Jamaica, Surinama " (LINN, syn. !)

5. SPERGULA
L., Sp. Pl. 440 (1753) & Gen. Pl., ed. 5, 199 (1754)

Annual herbs. Leaves linear, opposite but, from the axillary tufts of leaves, giving the superficial appearance of being in whorls ; stipules small, scarious, deciduous. Flowers in loose terminal dichasia, the dichasial structure reduced in some branches, hermaphrodite, pentamerous, pedicels deflexed after fertilization but later erect. Sepals free. Petals white, entire. Ovary unilocular with 3 or 5 free styles. Capsule deeply split into 3 or 5 valves.

S. arvensis *L.*, Sp. Pl. 440 (1753) ; F.T.A. 1 : 143 (1868) ; F.C.B. 2 : 138 (1951). Type : Europe (LINN, holo. !)

Erect, branched, stems up to 6 dm. or more in height, glabrous below, often glandular-viscid in the upper and inflorescence parts. Leaves 1–6·5 cm. long, 0·5–0·75 mm. broad, rounded or subacute, channelled on abaxial surface. Inflorescences 9–many-flowered. Sepals elliptic, obtuse, 4 mm. long, with white membranous margins. Petals obovate, 3·75 mm. long. Capsule with 20–25 seeds. Seeds biconvex, approximately circular in flat outline, with a narrow but not winged margin, black, with small pale club-shaped papillae.

KENYA. Eldoret District : Oldoinyo Sapuk, 23 July 1951, *Greenway* 8517 !
TANGANYIKA. Lushoto District : W. Usambara Mts., Mkuzi, 17 April 1953, *Drummond & Hemsley* 2118 !
DISTR. K3, 4 ; T3 ; now almost cosmopolitan, often as an introduced weed
HAB. Weed of cultivated and waste places, and dry river beds ; 1600–2550 m.

6. POLLICHIA
Ait., Hort. Kew., ed. 1, 1 : 5 (1789) & 3 : 505 (1789)

Branched subshrubs with rather stiff branches. Leaves opposite but giving the appearance of being in whorls (falsely verticillate) ; stipules free, scarious. Flowers hermaphrodite, in dense subsessile axillary cymes ; receptacle cupuliform. Sepals 5, free. Petals 5, minute. Stamens 1–2, with the perianth parts perigynous. Ovary with 2 basal ovules and a filiform persistent style. Fruits one-seeded, indehiscent with persistent sepals and borne on enlarged and fleshy rhachides and with the lower parts of the bracts also enlarged and fleshy.

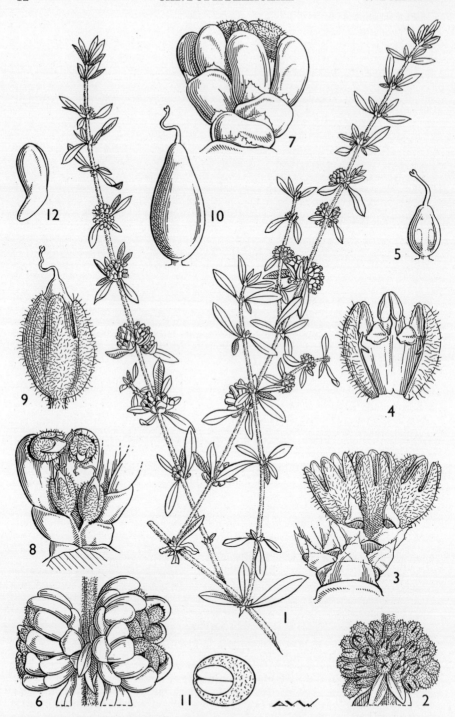

FIG. 5. *POLLICHIA CAMPESTRIS*, from *Gillett* 12816—**1,** part of plant, × 1 ; **2,** inflorescence, × 4 ; **3,** part of inflorescence, × 12 ; **4,** calyx opened to show petals and 1 stamen, × 18 ; **5,** gynoecium, × 18 ; **6,** infructescence, × 4 ; **7,** part of infructescence, outside view, × 6 ; **8,** part of infructescence, inside view, × 6 ; **9,** fruit with persistent calyx, × 18 ; **10,** fruit, × 18 ; **11,** transverse section of seed, × 18 ; **12,** embryo, × 18.

P. campestris *Ait.*, Hort. Kew., ed. 1, 1 : 5 (1789). Type : cult. from South Africa, Cape of Good Hope (BM, holo.!)

Woody in lower parts, much branched especially in the middle and upper parts, branches terete up to 7·5 dm. long, with persistent somewhat matted indumentum, white on the younger shoots ; internodes mostly 1–3 cm. long. Leaves narrowly oblanceolate or oblanceolate-linear, acute, apiculate, gradually narrowed below but no distinct petiole, 5–32 mm. long, 0·5–9 mm. broad, hairy when young (especially on the margins), glabrescent with increasing age ; stipules acuminate, 3–4 mm. long. Flowers 1·5 mm. long. Ovary papillose. Fig. 5.

UGANDA. Masaka District : Kabula, Sept. 1945, *Purseglove* 1832 !
KENYA. Naivasha District : Lake Naivasha, Apr. 1938, *Chandler* 2350 ! ; Machakos District : Sultan Hamud, 20 Sept. 1953, *Drummond & Hemsley* 4429 !
TANGANYIKA. Singida District : near Singida Boma, 3 Mar. 1928, *B. D. Burtt* 1363 !
DISTR. U1, 4 ; K3, 4, 6 ; T2, 5 ; widely distributed from Ethiopia to Cape Province, also in Arabia
HAB. Grassland, bushland, open woodland, and waste places, 1000–2340 m.

The development of the infructescences would repay detailed investigation. When ripe they are red to crimson and, very superficially, resemble raspberries. They are edible. The relationship of filaments to " petals " also requires investigation.

7. CORRIGIOLA

L., Sp. Pl. 271 (1753) & Gen. Pl., ed. 5, 132 (1754)

Annual, biennial, or perennial herbs with decumbent stems. Leaves spirally arranged, linear to narrowly elliptic or obovate, stipulate. Flowers small, in axillary and terminal clusters, often aggregated at the ends of the stem branches, hermaphrodite, pentamerous, with slightly perigynous receptacles. Sepals persistent, green with white margins. Petals shorter than or slightly exceeding the sepals in length. Stamens 5. Ovary unilocular, with a solitary basal ovule on a long funicle ; three subsessile stigmata. Fruits indehiscent, one-seeded, and enclosed in the persistent calyx.

C. littoralis *L.*, Sp. Pl. 271 (1753) subsp. **africana** *Turrill* in K.B. 1954 : 413. Type : Kenya, Kiambu District, Muguga, June 1952, *Verdcourt* 663 (K, holo. !)

A perennial (or sometimes biennial) herb, very much branched from the base, with slender prostrate branches 1–4·5 dm. long and internodes 0·5–1·0 cm. long. Leaves linear-oblanceolate or linear-elliptic, acute at the apex, with a gradually narrowed base, 0·7–2·0 cm. long, 1·5–3 mm. broad, more or less glaucous, with slightly prominent midrib but inconspicuous lateral nerves ; stipules membranaceous, unequally ovate, 3 mm. long, 2 mm. broad. Inflorescences compact, sessile or shortly pedunculate, on leafy branches ; pedicels up to 1 mm. long. Sepals ovate-oblong, obtuse or subobtuse, 1·25 mm. long in flower enlarging to 2 mm. long in fruit. Petals oblong, 0·75 mm. long, white. Fruit trigono-ellipsoid, 1·5 mm. tall. Fig. 6.

KENYA. Kiambu District : Limuru, 20 June 1941, *Wimbush* 1504 !
TANGANYIKA. Lushoto District : Shagai Forest, near Sunga, 17 May 1953, *Drummond & Hemsley* 2571 !
DISTR. K3, 4 ; T2, 3 ; Ethiopia, Eritrea
HAB. Waste places, ruderal, weed of arable land, 1200–2190 m.

Corrigiola paniculata *Peter*, F.D.O.-A. 2, app. p. 31 (1938) was based on *Peter* 34652 but no locality is given. Presumably, from the collector's number, the material was collected in Tanganyika in Tabora District near Malongwe. From the description and figure (tab. 38, fig. 1) it would seem that *C. paniculata* Peter is a synonym of *C. dryma-*

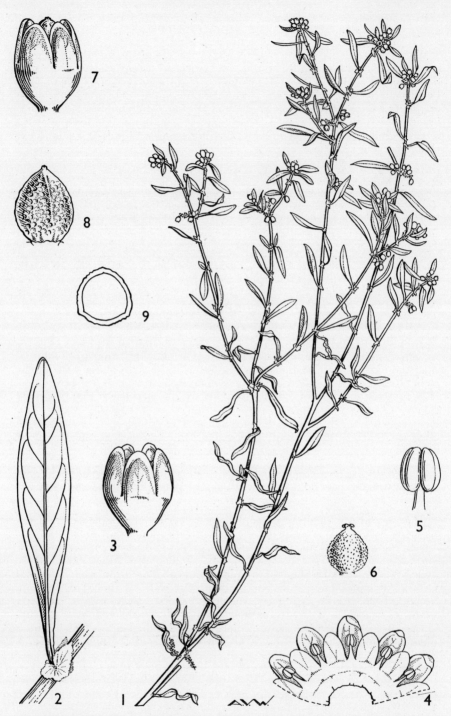

FIG. 6. *CORRIGIOLA LITTORALIS*, subsp. *AFRICANA*, from *Verdcourt* 663—**1**, part of plant, × 1 ; **2**, leaf showing stipules, × 4 ; **3**, flower, × 12 ; **4**, flower opened to show petals and stamens, × 12 ; **5**, anther, × 40 ; **6**, gynoecium, × 12 ; **7**, persistent calyx and fruit, × 12 ; **8**, fruit, × 12 ; **9**, transverse section of fruit, × 12.

rioides E. G. Baker in J.L.S. 40 : 181 (1911). This species, originally described from the Chimanimani Mountains, on the Southern Rhodesian and P.E.A. boundary, is known also from Nyasaland and various localities in Southern Rhodesia. Its range may well extend into Tanganyika. Dr. E. Potztal of Berlin-Dahlem has informed us that no material of *C. paniculata* has been found there and none has been seen by the present writer.

8. SAGINA

L., Sp. Pl. 128 (1753) & Gen. Pl., ed. 5, 62 (1754)

Tufted, sometimes cushion-like, or procumbent herbs. Leaves narrowly linear or subulate, exstipulate. Flowers small and relatively inconspicuous. Sepals 4–6, green. Petals as many as sepals and alternating with them, or absent ; when present white, often small and shorter than the sepals. Stamens usually 5–10 but the number may vary. Ovary unilocular, with an indefinite number of ovules ; styles as many as the sepals and alternating with them. Capsule opening to the base into as many valves as there are styles.

Sepals lanceolate to ovate-lanceolate, 2·3–4·8 mm. long ;
    petals 2 mm. long . . . . . . 1. *S. abyssinica*
Sepals ovate, 1·7–3·3 mm. long ; petals absent . . 2. *S. afroalpina*

1. **S. abyssinica** [*Hochst. ex*] *A. Rich.*, Tent. Fl. Abyss. 1 : 47 (1847) ; F.T.A. 1 : 142 (1868) ; F.W.T.A. 1 : 110 (1927) & ed. 2, 1 : 130 (1954) ; F.C.B. 2, 132 (1951) ; Svensk Bot. Tid. 48 : 208 (1954). Type : Ethiopia, Semen, near Demerki, *Schimper* 1148 (K, iso. !)

subsp. **aequinoctialis** *Hedberg* in Svensk Bot. Tid. 48 : 208 (1954). Type : Tanganyika, Kilimanjaro, above Peter's Hut, *Hedberg* 1412 (UPS, holo.)

A perennial glabrous herb varying considerably in habit according to the habitat conditions ; much branched, sometimes the branches very short and compact forming a close cushion, sometimes the flowering branches slender, more or less prostrate, and elongated, up to 2·4 dm. long. Leaves narrowly linear or somewhat subulate, acute, mucronulate, mostly 1–3 cm. long. Flowers in tufted forms immersed amongst the leaves of the cushions and on short pedicels, on elongated stems often with slender pedicels up to 2 cm. long. Sepals 4–6, ovate-lanceolate to lanceolate, adaxially concave, acute or subacute, 2·3–4·8 mm. long. Petals 4–6, inconspicuous, oblong-lanceolate, 2 mm. long. Capsule not protruding beyond the persistent sepals, 0·34–0·49 mm. broad. Seeds asymmetrically reniform, very minutely but distinctly papillose or tuberculate. Fig. 7.

UGANDA. Kigezi District : Mt. Mgahinga, June 1951, *Purseglove* 3708 !
KENYA. Aberdare Mts., Kinangop, below Karati falls, Apr. 1933, *Mrs. Albrechtsen* 2729 *in C.M.* 5164 !
TANGANYIKA. Kilimanjaro, Bismarck Hill to Peter's Hut, 28 Feb. 1954, *Greenway* 3922 !
DISTR. U1–3 ; K3, 4 ; T2 ; Ethiopia, Belgian Congo, Cameroons, Fernando Po
HAB. River banks and upland moor, 2100–4250 m.

2. **S. afroalpina** *Hedberg* in Svensk Bot. Tid. 48 : 209 (1954). Type : Tanganyika, Kilimanjaro, W. slope of Mawenzi, *Hedberg* 1250 (UPS, holo.)

A perennial glabrous herb with a more or less tufted and often cushion-like habit, the branches up to 4 or 6 cm. long. Leaves narrowly linear or somewhat subulate, acute, mucronulate, 0·4–1 cm. long. Flowers most often embedded in the cushions amongst the leafy branches and on short or very short pedicels, when the branches are somewhat elongated the pedicels are generally longer and up to 1·6 cm. long. Sepals 4–5, ovate, adaxially concave, rounded at apex, 1·7–3·3 mm. long. Petals absent (in the flowers

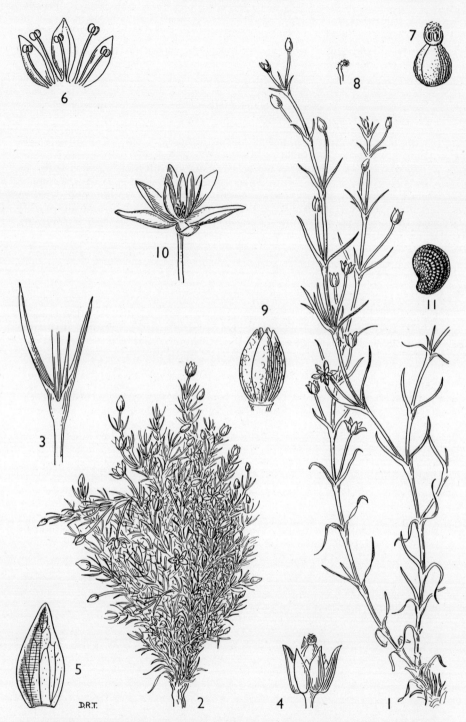

FIG. 7. *SAGINA ABYSSINICA* subsp. *AEQUINOCTIALIS*—**1** & **2**, parts of two plants to show habit, × 1 ; **3**, node, showing leaves, × 2 ; **4**, flower, × 4 ; **5**, sepal, × 6 ; **6**, part of flower from inside, showing arrangement of petals and stamens, × 9 ; **7**, gynoecium, × 9 ; **8**, style and stigma, × 9 ; **9**, capsule, with calyx removed, × 6 ; **10**, dehisced capsule surrounded by persistent sepals, × 4 ; **11**, seed, × 24. **1, 3, 4–11**, from *C. G. Rogers* 332 ; **2**, from *Purseglove* 779.

dissected). Capsule equalling but not exceeding the persistent sepals. Seeds somewhat irregular in outline but often more or less asymmetrically reniform, 0·5–0·65 mm. broad, almost smooth or very slightly rugulose or reticulated, more or less shining.

UGANDA. Mbale District : Bugishu, Sasa Camp, 16 Apr. 1950, *Mrs. L. M. Forbes* 268 !
KENYA. Mt. Kenya, 14 June 1933, *C. G. Rogers* 455 !
DISTR. U2, 3 ; K4 ; Ethiopia, Belgian Congo
HAB. Bogs and swamps of upland moor, 3150–4600 m.

SYN. *S. abyssinica* f. *apetala* Hauman in Bull. Acad. Roy. Belg., Cl. Sc., sér. 5, 19 : 704 (1933) ; F.C.B. 2, 132 (1951)

This species is very close to *S. brachysepala* Chiov. from Ethiopia. It differs mainly in the somewhat stouter pedicels and in the capsules not protruding.

9. MINUARTIA

L., Sp. Pl. 89 (1753) & Gen. Pl., ed. 5, 39 (1754) ; Hiern in J.B. 37 : 320 (1899)

Annual or perennial herbs or even subshrubs, with simple or branched stems. Leaves very variable in shape and texture, often acicular, subulate, or setaceous and three- (up to seven-) nerved, exstipulate. Sepals free, 5 (rarely 4), spreading at anthesis. Petals 5 (rarely 4), or absent, when present generally quite entire or rarely emarginate, contracted to base into a very short claw. Stamens typically 10 but sometimes reduced in number, provided with glands at the base of the filaments. Ovary superior, rarely semi-inferior at anthesis, unilocular ; styles usually 3. Capsule with the same number of valves as styles were present in the gynoecium, that is usually 3.

M. filifolia (*Forsk.*) *Mattfeld* in F.R. Beih. 15 : 93 (1922) ; Andrews, Fl. Pl. A.-E. Sudan 1 : 86 (1950). Type : Arabia, on Mt. Boka [about 40° 45′ N., 43° 40′ E.] (C, holo., cf. Mattfeld l.c.)

A perennial plant, subshrubby at the base, very much branched with branches that are more or less woody below and slender and herbaceous in the upper part, up to 2·5 dm. in total length, with internodes very short and shortly pubescent in the young shoots, elongating and becoming glabrous with age. Leaves opposite, especially aggregated towards the ends of the herbaceous branches or tufted on short axillary shoots, subulate-acicular but usually curved outwards, acute, somewhat widened and connate at the base, 5–10 mm. long, 0·1–0·2 mm. wide near the middle, glabrous. Inflorescences terminal on the branches, 1–6-flowered, pedicels 3–6 mm. long, glandular-pubescent. Sepals narrowly lanceolate, acuminate, 6 mm. long, 3-nerved, glabrous except slightly glandular-pubescent at base on outside. Petals about as long as sepals, white. Ovary triangular-ovoid with three styles. Capsule with 3 valves. Fig. 8.

TANGANYIKA. Mbulu District : Mt. Hanang, 2 Sept. 1932, *B. D. Burtt* 4040 !
DISTR. T2 ; Ethiopia, A.-E. Sudan, Somaliland, Eritrea
HAB. Crevices and cracks on exposed lava cliffs ; in upland moor, 3670 m.

SYN. *Arenaria filifolia* Forsk., Fl. Aegypt.-Arab. 211 (1775)
Alsine filifolia (Forsk.) Schweinf. in Bull. Herb. Boiss. 4, app. 2 : 175 (1896)
Alsine schimperi [Hochst. ex] A. Rich., Tent. Fl. Abyss. 1, 47 (1847). Type : Ethiopia, Tigré, Mt. Kubbi near Adowa, *Schimper* 549 (K, iso. !)
Arenaria schimperi (A. Rich.) Oliv. in F.T.A. 1 : 142 (1868) ; Fernald in Rhodora 21 : 6 (1919)
Minuartia schimperi (A. Rich.) Chiov. in Nuovo Giorn. Bot. Ital. 26 : 150 (1919)

VARIATION. There is some variation amongst the specimens of the species from Ethiopia and neighbouring countries. The Tanganyika material appears to come within the range of this variation, though, on the whole, it has the sepals somewhat

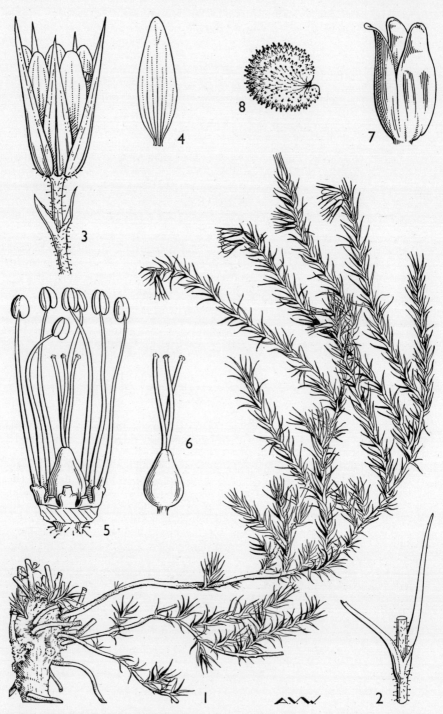

FIG. 8. *MINUARTIA FILIFOLIA*, from *Burtt* 4040—**1**, part of plant, × 1 ; **2**, node, × 4 ; **3**, flower, × 6 ; **4**, petal, × 6 ; **5**, stamens and gynoecium, 3 filaments cut, × 12 ; **6**, gynoecium, × 12 ; **7**, dehisced capsule, × 8 ; **8**, seed, × 24.

more acuminate and the capsules slightly shorter in length relative to the length of the sepals. With regard to indumentum the Tanganyika material has to be placed in var. *schimperi* (Hochst.) Schweinf. l.c.

10. CERASTIUM

L., Sp. Pl. 437 (1753) & Gen. Pl., ed. 5, 199 (1754)

Annual or perennial herbs or, rarely, subshrubs. Stems simple or branched. Leaves sessile, flat or rarely subulate, entire, exstipulate. Inflorescences terminal, dichotomous, often umbelliform occasionally with solitary flowers ; bracts herbaceous or membranous. Flowers hermaphrodite, tetramerous or pentamerous. Sepals free, with membranous margins. Petals white, emarginate or bifid, sometimes absent. Stamens 5 + 5, or fewer by reduction. Styles usually 4–5, ovules numerous. Capsule cylindric, often curved, longer than the sepals, with 8 or 10 teeth which are equal, short, erect or revolute.

Leaves up to 65 mm. long and 33 mm. broad ; flowers
    pentamerous ; petals 5·5–8 mm. long ; capsule
    straight and often not projecting beyond the
    persistent calyx, teeth revolute when ripe and dry   1. *C. indicum*
Leaves up to 35 mm. long and 10 mm. broad ; flowers
    pentamerous or tetramerous ; petals 5–11 mm.
    long ; capsule curved and projecting beyond the
    persistent calyx :
Petals equal in length to or shorter than the sepals
    or (rarely) absent ; capsule teeth straight   .   2. *C. octandrum*
Petals longer than the sepals ; capsule teeth more
    or less revolute when ripe and dry .   .   3. *C. afromontanum*

1. **C. indicum** *Wight & Arn.*, Prodr. Fl. Penins. Ind. Or. 1 : 43 (1834) ; Möschl in Mem. Soc. Brot. 7 : 53 (1951) ; F.W.T.A., ed. 2, 1 : 129 (1954). Type : India, Nilgirry Hills, *Wight* 149 (K, holo. !)

Perennial, pilose and glandular herb, with spreading and ascending slender branches, stems 20–60 cm. long. Leaves sessile or the lower ones with a very short petiole, narrowly lanceolate, lanceolate, or oblong-elliptic, acute and mucronate to shortly acuminate, 10–65 mm. long, 3–23 mm. broad, with silky hairs on both surfaces but more numerous on the abaxial surface. Inflorescence terminal with 3–13 flowers ; lowest bracts often foliaceous, the rest lanceolate-linear to linear and gradually smaller and more reduced. Flowers pentamerous. Sepals glandular but glabrescent towards the acute or subacute apex, inner ones scariose at the margins, 4–6 mm. long. Petals longer than the sepals, 5·5–8 mm. long, white. Capsule straight, often not projecting beyond the persistent calyx.

UGANDA. Kigezi District : Kachwekano Farm, Dec. 1949, *Purseglove* 3165 !
KENYA. Teita Hills, Yale peak, 13 Sept. 1953, *Drummond & Hemsley* 4283 !
TANGANYIKA. Lushoto District : W. Usambara Mts., Mkuzi, 19 Aug. 1950, *Verdcourt* 318 !
DISTR. U2, 3 ; K3–7 ; T2, 3, 6, 7 ; widely ranging in tropical Africa and southern India
HAB. Grassland, bushland, forest margins and glades, roadsides, pathsides, weed of cultivated land, 1050–2900 m.

SYN. *Arenaria africana* Hook. f. in J.L.S. 7 : 184 (1864). Type : Cameroon Mt., *Mann* 1941 (K, holo. !)
    Cerastium africanum (Hook. f.) Oliv. F.T.A. 1 : 141 (1868) ; F.W.T.A. 1, 110 (1927) ; F.C.B. 2 : 135 (1951)
    Stellaria schimperi Engl. in Abh. Preuss. Akad. Wiss. 1891 (2) : 212 (1892). Type : Ethiopia, Begemdir, near Debra-Tabor, *Schimper* 1383 (K. iso. !)
    C. schimperi (Engl.) De Wild., Pl. Bequaert. 1 : 204 (1922)

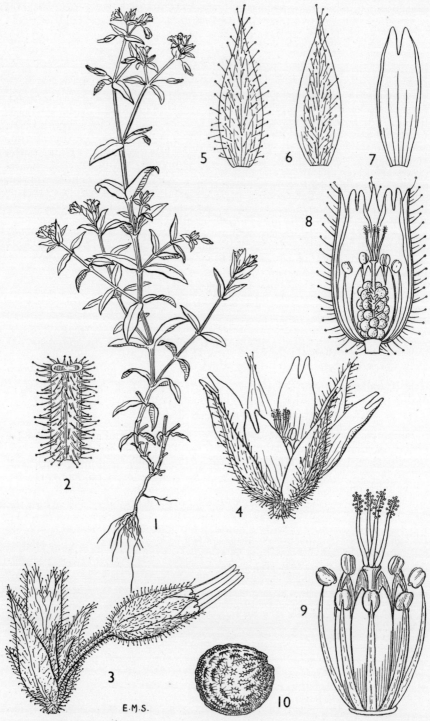

E.M.S.

Fig. 9. *CERASTIUM OCTANDRUM*, from *A. S. Thomas* 306—**1**, part of plant, × ½ ; **2**, part of stem showing glandular hairs, × 6 ; **3**, part of inflorescence, showing flower and fruit, × 3 ; **4**, flower, × 6 ; **5 & 6**, sepals, × 6 ; **7**, petal, × 6 ; **8**, sectional view of flower, × 6 ; **9**, stamens and gynoecium, × 12 ; **10**, seed, × 36,

Variation.   *C. indicum* appears in East Africa to be less variable, except in leaf size, than some other species.   The var. *ruwenzoriense* (Williams) Möschl (l.c.) is not taxonomically distinguishable from the type and should be named var. *indicum*.

2. **C. octandrum** [*Hochst. ex*] *A. Rich.*, Tent. Fl. Abyss. 1 :   45 (1847) ; Möschl in Mem. Soc. Brot. 7 :   24 (1951) ; F.W.T.A., ed. 2, 1 :   129 (1954). Type :   Ethiopia, Tigré, near Adowa, *Schimper* 1841 (K, iso. !)

Annual, more or less pilose and glandular herb, very plastic, sometimes with rather compact habit but usually with spreading and ascending branches ;   stems 2–42 cm. long.   Leaves sessile, the lower spathulate, the upper ovate, ovate-oblong, or elliptic, subacute and apiculate or very shortly acuminate, very variable in size, 3–25 mm. long, 1–10 mm. broad, commonly pilose on both surfaces but variable in density of hairs.   Inflorescences terminal, with 1–59 flowers ;   bracts foliaceous, not scariose.   Flowers tetramerous or pentamerous.   Sepals pilose and glandular or glabrescent, scariose at the margins, 4–7 mm. long.   Petals equal in length to or shorter than the sepals, white, rarely absent.   Stamens 8–10.   Capsules subcylindrical, somewhat incurved, 4–13 mm. long, projecting beyond the persistent calyx when mature.   Fig. 9.

Uganda.   Mbale District :   Bugishu, Bulago, 27 Aug. 1932, *A. S. Thomas* 306 !
Kenya.   Nakuru District :   near Molo, 24 July 1951, *Bogdan* 3176 ! ;   Mt. Kenya, 19 March 1922, *Fries* 1326 (apetalous) !
Tanganyika.   Moshi District :   Kilimanjaro, Shiva Plateau, Feb. 1928, *A. E. Haarer* 1104 !
Distr.   U2, 3 ;   K1 (?), 3, 4 ;   T2, 7 ;   Eritrea, Ethiopia, A.-E. Sudan, Cameroons, Belgian Congo
Hab.   By streams or roadsides, in bogs, upland moor and bushland, 1920–4200 m.

Variation.   There is no doubt, that this species is extremely plastic and habit especially varies greatly with the habitat.   Möschl (l.c.) has two varieties var. *humile* (Braun) Möschl and var. *adnivale* (Chiovenda) Möschl.   The former should be known as var. *octandrum* according to the present International Rules of Nomenclature.   The latter is a reduction to varietal status of *C. adnivale* Chiovenda in Bull. Soc. Bot. Ital. 1917, 21. Type :   Uganda, Ruwenzori, Valley of the Lakes, 4500 m., June 1906, *Luigi di Savoia*.   The differences between the two varieties concern the degree of development and arrangement of the indumentum.   While specimens fitting well the descriptions given by Möschl have been examined, others are found to be intermediate.   The extreme differences are :
      var. *octandrum* (var. *humile*) :   peduncles and stems pilose over the whole surface ; the outermost sepal with eglandular and glandular hairs on the abaxial surface ; leaves pilose
      var. *adnivale* :   peduncles and stems with single lines of hairs or glabrescent ;   sepals and leaves glabrous or with few hairs
      Extreme plants named var. *adnivale* are from Uganda and the following is quoted as an example :   Kigezi District :   Mt. Muhavura, in crater, June 1939, *Purseglove* 753 !
      From the material examined, *C. keniense* T. C. E. Fries & Weimarck in Bot. Not. 1929, 290 (Type :   Mt. Kenya, W. side, 2–5 Feb. 1922, *Fries* 1377) appears to be *C. octandrum* var. *adnivale* and not a variety of *C. afromontanum* as it is made by Möschl.   Only *Fries* 1377a, in Herb. Kew., has been seen of the material quoted for *C. keniense.*

3. **C. afromontanum** *T. C. E. Fries & Weimarck* in Bot. Not. 1929, 294 ; Möschl in Mem. Soc. Brot. 7 :   41 (1951).   Type :   Kenya, Aberdare Mts., near Sattima, 19 Feb. 1922, *Fries* 2634 (UPS, holo., K, iso. !)

Perennial, more or less pilose and glandular herb, with usually long spreading and ascending slender branches ;   stems 4–40 cm. long.   Leaves sessile, ovate, oblong-elliptic, to narrowly oblong and almost linear, acute, variable in size as well as in shape, 4–35 mm. long, 2–9 mm. broad, pilose and glandular on both surfaces in varying degrees, especially on the midrib on the abaxial surface.   Inflorescences terminal, with 2–23 flowers ;   lower bracts similar to the foliage leaves, no abrupt transition from leaves to reduced bracts. Flowers tetramerous or pentamerous. Sepals glandular-pilose, scariose at the margins, 4–7 mm. long. Petals longer than the sepals, 8–11 mm. long, white. Capsule subcylindrical, 7–10 mm. long, when fully ripe well projecting beyond the persistent calyx and somewhat curved.

UGANDA. Mbale District : Bulambuli, Bugishu, 5 Sept. 1932, *A. S. Thomas* 560 !
KENYA. Nakura District : Thompson's Falls district, 23 Oct. 1931, *Pierce* 1480 !
TANGANYIKA. Mbulu District : Mt. Hanang, or Guruwe Mt., 26 Dec. 1929, *B. D. Burtt* 2269 !
DISTR. U1–3 ; K3–5 ; T2, 3, 7 ; A.-E. Sudan (Imatong Mts.)
HAB. Upland and moor grasslands, in upland moor and bushland, 2100–3540 m.

SYN. *C. africanum* Oliv. var. *kilimanjarensis* Williams in J.B. 36 : 342 (1898). Type : Tanganyika, Kilimanjaro, *Volkens* 792 (K, iso. !)
   C. caespitosum Gilib. var. *kilimandscharicum* Engl. in Urban & Graebner, Festschr. - - - Paul Ascherson, 566 (1904). Type : Tanganyika, Kilimanjaro, *Uhlig* 628 (B, lecto., EA, iso-lecto. !)
   C. africanum Oliv. var. *jaegeri* Engl. in E.J. 48 : 380 (1913), *sec.* Möschl. l.c. Type : Tanganyika, Mbulu District, crater-rim of Mt. Oldeani, *Jaeger* 403 (B, lecto.)
   C. aberdaricum T. C. E. Fries & Weimarck in Bot. Not. 1929, 291. Type : Kenya, Aberdare Mts., near Sattima, *Fries* 2410 (UPS, holo.)
   C. kilimandscharicum (Engl.) T. C. E. Fries & Weimarck l.c. 292
   C. pycnophyllum Peter, F.D.O.-A. 2, app. p. 30, t. 37/1 (1932). Type : Tanganyika, Kilimanjaro, *Peter*, 805 (B, holo. †)
   C. bambuseti (T. C. E. Fries & Weimarck) Weimarck in Svensk Bot. Tidskr. 27 : 413 (1933). Type : Mt. Kenya, *Fries* 1175 (UPS, holo.)

VARIATIONS. This species is undoubtedly variable but how much of the variation is plasticity in reaction to varying environments and how much is due to genetic differences is unknown. The vars. *granvikii* T. C. E. Fries & Weimarck, l.c. 294, *bambuseti* T. C. E. Fries & Weimarck, l.c. 295, and *kilimanjarense* (Williams) Möschl, l.c. 44, appear to have little or no taxonomic value. The var. *keniense* (T. C. E. Fries & Weimarck) Möschl l.c. 45 is referred to *C. octandrum* [Hochst. ex] A. Rich. var. *adnivale* (Chiov.) Möschl. (q.v.) *C. africanum* var. *jaegeri* Engl. is, from the description, to be referred to *C. afromontanum*.
   C. glomeratum *Thuill.* (*C. viscosum* auct. plur., nom. ambig.) has been recorded by several collectors from Tanganyika : Lushoto District, near Amani, as an adventive. It is morphologically related to *C. octandrum* [Hochst. ex] A. Rich. from which it differs especially in having the lower (and rarely all) flowers without petals (nearly always) ; petals with generally a ciliate claw ; sepals more glandular ; stamens often only 5 ; placenta not branched or radiate ; seeds 0·5 mm. in diameter. A standard specimen is : *Verdcourt* 301, in a drain around building with *Ageratum, Galinsoga, Chloris, Oxalis, Pilea,* and *Drymaria,* 1 Aug. 1950 !

11. ARENARIA

L., Sp. Pl. 423 (1753) & Gen. Pl., ed. 5, 193 (1754)

Herbs or, rarely, subshrubs, mostly small and often prostrate, tufted, or cushion-like in habit. Leaves opposite, ovate, lanceolate, or linear, exstipulate. Flowers solitary or in dichasial cymes, usually pentamerous. Sepals free. Petals entire or slightly emarginate or dentate, white or pink. Stamens 10 or fewer by reduction. Ovary unilocular ; styles free, usually 3. Capsule opening by 6 (rarely 8 or 10) teeth. Seeds several to numerous, reniform.

A. foliacea *Turrill* in K.B. 1954 : 415. Type : Tanganyika, Lushoto District, Shume Forest, *Drummond & Hemsley* 2692 (K, holo. !)

An annual herb, branched often from the base, with slender stems 4–7 cm. long and leafy to the apex, internodes 4–20 mm. long, glandular hispidulous. Leaves oblanceolate or narrowly elliptic-oblanceolate, acute or shortly acuminate, gradually narrowed at the base but not distinctly petiolate, 3–11 mm. long, 1–4 mm. broad, with inconspicuous nervation, glandular ciliate at the margin. Flowers solitary, on slender pedicels 2–6 mm. long, glandular hispidulous. Sepals lanceolate, acuminate, 3–4 mm. long. Petals small, white. Capsule 2·75 mm. long, not strongly flask-shaped. Fig. 10.

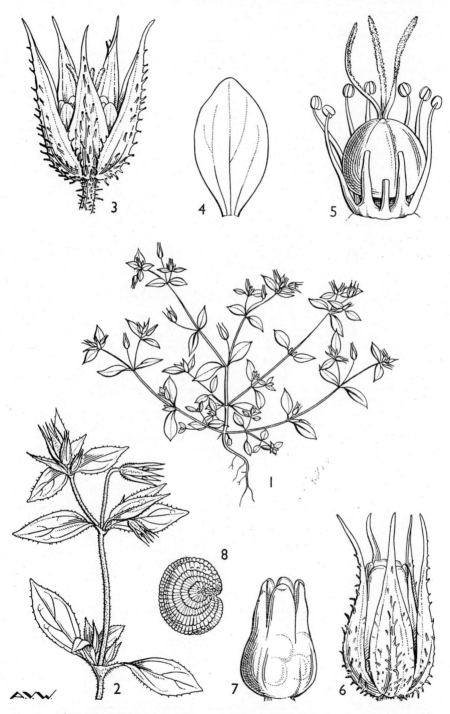

FIG. 10.   *ARENARIA FOLIACEA*, from *Drummond & Hemsley* 2692—**1**, plant, × 1 ; **2**, flowering branch, × 3 ; **3**, flower, × 12 ; **4**, petal, × 24 ; **5**, stamens and gynoecium, 3 filaments cut, × 24 ; **6**, persistent calyx and capsule, × 12 ; **7**, dehiscing capsule, × 12 ; **8**, seed, × 40.

TANGANYIKA. Lushoto District : West Usambara Mts., Shume Forest, about 3 km.
    SE. of Manolo, 23 May 1953, *Drummond & Hemsley* 2692 !
DISTR. **T3** ; Somaliland Protectorate (?)
HAB. Open grassy area at fringe of dry evergreen forest, 1800 m.

12. STELLARIA

L., Sp. Pl. 421 (1753) & Gen. Pl., ed. 5, 193 (1754)

Annual or perennial herbs, often with slender diffuse stems, glabrous or
hairy. Leaves opposite, simple, entire, flat to (rarely) subulate, exstipulate.
Inflorescences dichasial cymes which are terminal to the branches, rarely
the flowers solitary. Sepals free, 4–5. Petals 4–5, more or less deeply
bilobed, white, or occasionally absent. Stamens 5 + 5, or fewer by reduction. Styles generally 3 but sometimes 2, free. Capsule unilocular with 3
(or rarely 2) or 6 (rarely 4) valves. Seeds either numerous or 1–4.

Sepals 5, styles 3 :
  Internodes with a longitudinal line of hairs, changing
    sides at the nodes . . . .   1. *S. media*
  Upper internodes with glandular hairs all round but the
    lower more or less glabrescent . . . .   2. *S. mannii*
Sepals 4, styles 2 . . . .   3. *S. sennii*

1. **S. media** (*L.*) *Vill.* Hist. Pl. Dauph. 3 : 615 (1789). Type : Europe
(LINN, holo. !)

An annual herb with diffuse leafy stems. Leaves very variable in size
and shape, the uppermost usually sessile and the lower stalked, blades
mostly ovate or elliptic, acute or shortly acuminate, glabrous or ciliate
near the base. Flowers in often leafy dichasia. Sepals 3·5–5 mm. long,
with a narrow membranous margin. Petals shorter than the sepals, deeply
bifid. Stamens 3–10. Capsules more or less exceeding the calyx in length,
with 6 valves, the fruiting pedicels usually turning downwards at some
stage of ripening of the capsules. Seeds about 1 mm. in breadth, tubercled.

KENYA. Naivasha District : 24 km. N. of Gilgil, 9 Sept. 1948, *Bogdan* 2008 !
TANGANYIKA. Lushoto District : Mkuzi, by Mkusu River, 19 Aug. 1950. *Verdcourt* 323 !
DISTR. **K3** ; **T**2, 3, 7 ; now practically cosmopolitan as a weed or ruderal
HAB. A weed of cultivated land, or a ruderal ; 1290–2370 m.

SYN. *Alsine media* L., Sp. Pl. 272 (1753)

VARIATION. This species, presumably introduced as a weed into East Africa, is very
    variable. Not only is it extremely plastic according to the environmental factors but
    there are also variants that have, or probably have, a genetic basis different from that
    of the type. Taxonomists have expressed different opinions as to the best status to
    be accorded to some of the taxa of this section of the genus. It should be noted that
    one sheet (*Geilinger* 2205) from Tanganyika, Rungwe, shows some characters (as habit
    and flower size) of *Stellaria apetala* Ucria, which is often considered a variant of *S.
    media* (L.) Vill. The flowers, however, have petals 1·75 mm. long and seeds 1 mm.
    wide.

2. **S. mannii** *Hook. f.* in J.L.S. 7 : 183 (1864) ; F.T.A. 1 : 141 (1868) ;
F.W.T.A. 1 : 110 (1927) & ed. 2, 1 : 129 (1954). Type : Cameroon Moun-
tain, *Mann* 1940 (K, holo. !)

Stems procumbent to weakly ascending, sometimes rooting at some nodes,
glabrescent with scattered hairs to glabrous in the lower parts or with hairs
retained in the furrows. Leaves all petiolate ; blades ovate, acute and
apiculate or shortly acuminate, rounded or slightly cuneate at the base,
when well formed 1·5–4·5 cm. long and 1–3 cm. broad, midrib moderately
conspicuous, lateral veins rather obscure but all anastomosing very close
to the margin, with scattered and often glandular hairs ; petioles 0·4–3 cm.

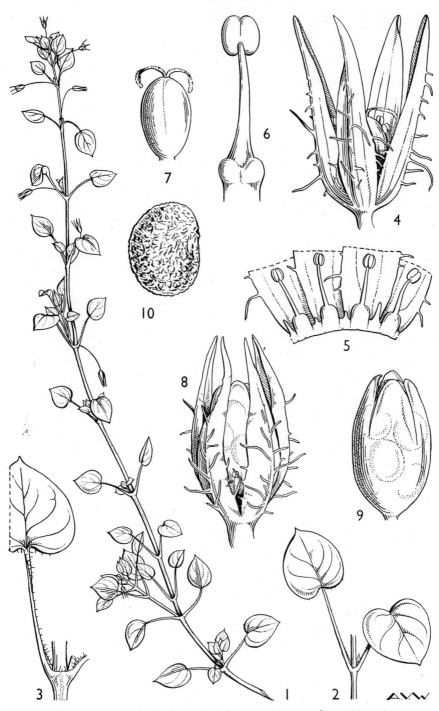

Fig. 11. *STELLARIA SENNII*, from *Tweedie* 1071—**1,** part of plant, × 1 ; **2,** pair of larger leaves, × 1 ;
3, node showing indumentum of petiole, × 2 ; **4,** flower, × 12 ; **5,** flower opened out to show vestigial
petals and stamens (upper pair of sepals and gynoecium removed), × 12 ; **6,** stamen viewed from back,
× 36 ; **7,** gynoecium, × 12 ; **8,** persistent calyx enclosing capsule, × 8 ; **9,** dehiscing capsule, × 8 ;
10, seed, × 18.

long, slightly amplexicaul at the base and there usually more or less connate, glandular-hirsute. Inflorescence terminal to the main branches, densely but shortly glandular-pubescent ; pedicels at first very short but elongating to 8 mm. Sepals ovate-lanceolate, acute or shortly acuminate, 4·5–6 mm. long, glandular on the abaxial surface. Petals strongly bifid, often about equal in length to or slightly shorter than the sepals or up to half as long again as the sepals, white. Stamens 10, when all fertile. Styles 3.

UGANDA. Kigezi District : Kinaba, Luhiza, Mar. 1947, *Purseglove* 2352 !
KENYA. Machakos District : Chyulu Hills, 11 June 1938, *Bally* 7632 !
TANGANYIKA. Lushoto District : Lutindi, July 1893, *Holst* 3275 !
DISTR. U2 ; K4 ; T3, 7 ; Ethiopia, Cameroons, Fernando Po, S. Tomé, Southern Rhodesia, Nyasaland, Belgian Congo, Madagascar
HAB. Forest floor and coffee plantations, 1650–2400 m.

VARIATION. The specimens seen and accepted as *S. mannii* show some variation in the degree of development of the glandular indumentum and in some floral characters such as apex of sepals and length of petals. The name *Stellaria rugegensis* is used in F.P.N.A. 1 : 157 (1948) and in F.C.B. 2 : 133 (1951) but this is a *nomen nudum* in E.J. 48 : 380 (1912) and, so far as traced, has never had a Latin description attached to it. One supposes from the available evidence that *S. mannii* is the equivalent of *S. rugegensis* f. *parvipetala* Balle, F.C.B. 2 : 134 (1951) while typical *S. rugegensis*, had it been validly published, would be reduced to a variant of *S. mannii*. It is a matter for further study, with more and better material, field observations, and genetical experiments, to determine how far paramorphs (i.e. variants within the species) should be distinguished. There are plants with female (male-sterile) flowers and the range of petal size is worth detailed investigation.

3. **S. sennii** *Chiov.* in Atti R. Accad. Ital., Mem. Cl. Sc. Fis. etc. 11, 20 (1940). Type : Ethiopia, Scioa, Bosco Mannagascià, *Senni* 2165 (FI, holo. !)

Stems procumbent to weakly ascending, diffuse, generally glabrous or with one or two rows of hairs. Leaves all distinctly petiolate ; blades ovate to broadly ovate, acute, shortly apiculate, more or less cordate at the base, varying considerably in size, from 3 mm. long and 3 mm. broad to 18 mm. long and 14 mm. broad, midrib and main veins fairly conspicuous, the latter anastomosing near the margin, glabrous or with a few scattered hairs ; petioles 0·3–1·7 cm. long, pilose near the base and with scattered hairs or glabrous in the upper part. Flowers solitary and axillary to the upper leaves ; pedicels 5–18 mm. long, with some scattered hairs or glabrous. Sepals 4, narrowly lanceolate, acute or shortly acuminate, 3–4 mm. long, pilose on abaxial surface, margins scarious. Petals absent (at least usually) or vestigial. Styles 2. Capsule shorter than the calyx. Seeds 1–4 per capsule, papillate. Fig. 11.

UGANDA. Kigezi District : Mt. Mgahinga, June 1949, *Purseglove* 2933 !
KENYA. Ravine District ; 2nd day's march [W.] from Eldama Ravine, *Whyte* !
TANGANYIKA. Moshi District : Useri, Jan. 1929, *Haarer* 1790 ; Morogoro District : Uluguru Mts., Tanana, 24 Jan. 1935, *E. M. Bruce* 635 !
DISTR. U2, 3 ; K3, 4 ; T2, 6, 7 ; Ethiopia, Belgian Congo, Nyasaland
HAB. Bushland, forest, roadsides, 1500–3300 m.

SYN. *S. media* L. var. *brauniana* [Fenzl ex] Engl. P.O.A. C : 176 (1895)
 S. brauniana [Engl. ex] Robyns F.P.N.A. 1 : 158 (1948) ; F.C.B. 2 : 134 (1951) ; Hedberg in Svensk Bot. Tid. 48 : 201 (1954). Type : Ethiopia, Semen, in Maschiha Valley near Maua, *Schimper* (B, lecto †)

13. UEBELINIA

Hochst. in Flora 24 : 664 (1841)

Low growing, procumbent or more or less erect or ascending herbs. Leaves usually narrowed into very short petioles which may be slightly connate at the base. Calyx gamosepalous, with 5 or 4 lobes (sepals), with 10 or 8 main

Fig. 12.   *UEBELINIA KIGESIENSIS*, from *Eggeling* 958—**1**, part of plant, × 1 ; **2**, leaf, lower surface,
× 2 ; **3**, flower, × 6 ; **4**, flower, with part removed to show petals, stamens and gynoecium, × 6 ; **5**, part
of flower from inside, showing arrangement of petals and stamens, × 6 ; **6**, petal, × 6 ; **7**, stamen, × 12 ;
8, gynoecium, × 6 ; **9**, dehisced capsule, × 6 ; **10**, seed × 12.

nerves. Petals 5 or 4 narrowly or broadly spathuliform, longer or shorter than the calyx, white. Stamens 4–10. Ovary unilocular, with 3–5 free styles, alternating with the sepals when equal in number to them ; ovules 1–9. Capsule septicidal, included in the persistent calyx, with 1–7 seeds, opening by 3–5 teeth or valves.

In T. C. E. Fries's account of *Uebelinia*, in F.R. 19 : 81–92 (1923), a good deal of stress is laid on the numbers of the parts in the flower. This treatment is justified in part, but the numbers, especially of the stamens, are not always constant even in flowers of the same individual plant. There may occasionally be found some divergence from the numbers for the floral organs given in the descriptions below since these are necessarily based on dissections of a limited (and often small) number of flowers. Attention may be called to the exceptional interest of this genus, both for the flower structure and for the phytogeographical problems raised by the ranges of the species. Field botanists are asked to make extensive collecting and intensive population studies of the species of *Uebelinia*.

Styles 4–5 ; ovules and seeds 4–9 :
  Leaf-blades 1–2 cm. long :
    Pedicels 0·2–0·3 cm. long ; flowers 4–5 mm. long .   1. *U. abyssinica*
    Pedicels 0·7–1·5 cm. long ; flowers about 7 mm.
     long . . . .   2. *U. kigesiensis*
  Leaf-blades 1 cm. or less in length ; pedicels 1–2 mm.
    long ; flowers about 5 mm. long . . .   3. *U. rotundifolia*
Styles 3 ; ovule and seed solitary :
  Internodes glabrescent ; leaf-blades up to 1 cm. long ;
    pedicels 0·1–0·4 cm. long at anthesis ; flowers
    about 5 mm. long . . . . . .   4. *U. crassifolia*
  Internodes hirsute in the upper part ; leaf-blades up
    to 2 cm. long ; pedicels 1–2·5 cm. long at
    anthesis ; flowers about 7 mm. long . .   5. *U. kivuensis*

1. **U. abyssinica** *Hochst.* in Flora 24 : 665 (1841) ; F.T.A. 1 : 140 (1868), *pro parte* ; F.C.B. 2 : 151 (1951). Type : Ethiopia, Tigré, near Adowa, *Schimper* 302 (K, iso. !)

A procumbent herb with branching and elongating stems rooting at many of the lower nodes, internodes 1–3 cm. long, rather densely hirsute with the hairs all round the stem. Leaves nearly sessile or narrowed at the base, often rather suddenly, to a flattened petiole 1–2 mm. long ; blades elliptic or obovate-elliptic, mucronulate, 1–1·5 cm. long, setulose-ciliate at the margins, with scattered bristles especially on the midrib on the abaxial surface, usually glabrous on the adaxial surface. Pedicels 2–3 mm. long in Uganda material. Flowers 4–5 mm. long. Stamens 7–8 in Uganda material. Styles 5. Seeds 7 (9).

UGANDA. Mbale District : Bugishu, Bulago, 27 Aug. 1932, *A. S. Thomas* 305 !
DISTR. U3 ; Ethiopia, Ruanda
HAB. Forest clearings and secondary bushland, 1950 m.

The determination of the one collection (*A. S. Thomas* 305) available from the area of the present Flora as *U. abyssinica* is made with some hesitation. The internodes are somewhat more densely hirsute than are those of the type (*Schimper* 302), the stems are procumbent or scrambling and apparently not or scarcely ascending in the upper part, and the stamens (in the three flowers dissected) are 7 to 8 (not 10). Plants grown at Kew from Ethiopian seed (*H. Scott* 268), however, morphologically link up the Uganda material with Schimper's type. It is also clear from the results of dissecting many flowers that, in some at least of the species of this genus, the number of stamens is not always a constant character.

2. **U. kigesiensis** *R. Good* in J.B. 62 : 332 (1924).   Type : Uganda, Kigezi District, Behungi, *Godman* 237 (BM, holo. !)

A sub-erect, ascending, or partially procumbent herb with elongating and more or less branching stems apparently not rooting at the nodes or at most only at the lowest, internodes 1–6·5 cm. long, the youngest hirsute especially in the upper part and a tendency for the hairs to be in longitudinal lines, glabrescent to glabrous in the lower part of every internode and often throughout the lower internodes. Leaves nearly sessile or narrowed at the base to a flattened petiole 1–2 mm. long ; blades oblong-elliptic or ovate-elliptic, acute and mucronulate, 1–2 cm. long, setulose-ciliate at the margins, glabrous on both surfaces or with a few scattered bristles on the abaxial surface. Pedicels 0·7–1·5 cm. long (usually 1 cm. or less at anthesis), hirsute. Flowers about 7 mm. long. Stamens 10. Styles 5. Ovules 5–7. Fig. 12.

UGANDA. Kigezi District : Behungi swamp, 1 Dec. 1930, *B. D. Burtt* 2932 !
DISTR. **U**2 ; endemic in Kigezi District of Uganda
HAB. An aquatic or marsh plant ; 2100–2400 m.

3. **U. rotundifolia** *Oliver* in J.L.S. 21 : 397 (1885) & in Hook., Ic. Pl. 15, t. 1492 (1885) ; *T. C. E. Fries* in F.R. 19 : 88 (1923). Type : Tanganyika, Moshi District, Kilimanjaro, *J. Thomson* (K, holo. !)

A procumbent herb with branching often elongated stems rooting at many of the nodes, internodes 0·5–2·5 cm. long with reflexed hairs more or less in two longitudinal lines. Leaves rather thick, usually narrowed at the base to a flattened petiole 1–2 mm. long ; blades suborbicular or oblate, mostly mucronulate, 5–10 mm. long, setulose-ciliate at the margins, otherwise with a few scattered bristles or glabrous. Flowers about 5 mm. long, subsessile or with pedicels 1–2 mm. long. Stamens 9–10. Styles 4–5. Seeds 4–5.

TANGANYIKA. Kilimanjaro, Marangu, Sept. 1893, *Volkens* 972 !
DISTR. **T**2 ; endemic on Kilimanjaro
HAB. Forest and wet shady stream banks, 2600–2900 m.

4. **U. crassifolia** *T. C. E. Fries* in F.R. 19 : 91 (1923). Type : Mt. Kenya, *Fries* 1173 (UPS, holo., K, iso. !)

A procumbent herb with branching often elongated stems rooting at many of the nodes, internodes 0·5–2 (3) cm. long, generally glabrescent but sometimes with rather few hairs perhaps in two obscure longitudinal lines. Leaves rather thick, usually narrowed at the base to a flattened petiole 1–2 mm. long ; blades suborbicular or oblate, mostly mucronulate, 0·4–1 cm. long, setulose-ciliate at the margins, otherwise with a very few scattered bristles or glabrous. Flowers about 5 mm. long, subsessile or with pedicels 0·1–0·4 cm. long. Stamens 8. Styles 3. Seed solitary.

KENYA. Aberdare Mts., Kinangop, June 1931, *Miss Dent* 1306 !
DISTR. **K**3, 4 ; endemic on Mt. Kenya and Aberdare Mts.
HAB. Open grassy places in moist bamboo thicket and upland moorlands, 2500–3180 m.

5. **U. kivuensis** *T. C. E. Fries* in F.R. 19 : 90 (1923). Type : Ruanda-Urundi, Rugege Forest, Rukarara [Lukarara] R., *Mildbraed* 974 (B, holo. †)

A procumbent herb with branching and elongating stems rooting at many of the lower nodes, internodes 0·7–4 cm. long, hirsute in more or less clearly marked longitudinal lines in the upper part glabrescent to glabrous in the lower part of every internode. Leaves nearly sessile or narrowed at the base, often rather suddenly, to a flattened petiole 2–3 mm. long ; blades elliptic to nearly orbicular or obovate, mucronulate, 0·8–2 cm. long, setulose-ciliate at the margins, with scattered bristles on the abaxial surface, with shorter hairs especially on the midrib or glabrous or glabrescent on the adaxial surface. Flowers about 7 mm. long, with slender pedicels 1–2·5 cm. (in fruit 3 cm. or more) long, more or less hirsute. Stamens 8. Styles 3. Ovule solitary.

UGANDA. Kigezi District : Bufumbira, Nyarusiza, Jan. 1947, *Purseglove* 2297 !
DISTR. U2 ; Ruanda
HAB. Upland grassland, moorland, and moist bamboo thicket, 1800–2600 m.

14. MELANDRIUM

Roehl., Deutschl. Flora, ed. 2, vol. 2 : 37, 274 (1812)

Annual or perennial herbs, sometimes suffruticose. Leaves exstipulate, not or only slightly connate at the base. Inflorescences cymose, paniculate, secund, or reduced to a single flower ; no calycine bracts. Calyx tubular or dilated, 5 teeth, 8–10 main veins. Petals 5, with a long narrow claw and a bilobed limb. Flowers hermaphrodite or dioeciously unisexual. Stamens 5 + 5 in male or hermaphrodite flowers. Ovary (and capsule) unilocular ; styles 3 or 5, free. Capsules dehiscing by 6 or 10 teeth or valves.

Leaves narrowly oblanceolate or linear-lanceolate ;
    inflorescences secund .   .   .   .   .   1. *M. syngei*
Leaves oblong-elliptic ; inflorescences of loose few-
    flowered cymes .   .   .   .   .   2. *M. lomalasinense*

1. **M. syngei** *Turrill* in K.B. 1954 : 411. Type : Uganda, Elgon, May 1935, *Synge* S 1912. (BM, holo. !)

A perennial branched herb, with stems up to 2·2 dm. long and somewhat woody near the base, very leafy ; internodes with dense spreading pubescence. Leaves narrowly oblanceolate or linear-lanceolate, acute and shortly thickened at the apex, gradually narrowed to the base, 1·4–3·5 cm. long, 2–7 mm. broad, densely pubescent on both surfaces. Inflorescence secund, 2–6 cm. long, 2–5-flowered, densely pubescent ; the upper bracts narrowly lanceolate, 3–9 mm. long. Calyx 1·1–1·3 cm. long, 4–6 mm. in diameter, densely pubescent, with triangular-lanceolate teeth. Petals 1·0–1·2 cm. long, white, pubescent on abaxial surface, with well-developed coronal scales.

UGANDA. Mbale District : Elgon, NW. slopes, *Synge* S 1912 !
DISTR. U3 ; endemic on Elgon
HAB. Damp rocks and cliff face in upland moor, 3750 m.

2. **M. lomalasinense** *Engl.* in E.J. 48 : 383 (1912). Type : Tanganyika, Mbulu District, Mt. Loolmalasin, Feb. 1907, *Jaeger* 473 (B, holo. †)

A straggling herb, probably perennial. Stems erect, the internodes covered with somewhat stiff retrorse hairs. Leaves oblong-elliptic, acute or shortly acuminate, narrowed at the base into a very short petiole or sessile, 1·7–3 cm. long, 0·5–1·2 cm. broad, glabrous on adaxial surface, with somewhat stiff hairs on the abaxial surface especially on the midrib. Inflorescences terminal, of loose, few-flowered cymes, with the flowers on slender retrorsely hairy pedicels 0·5–2·5 cm. long. Calyx 9–10 mm. in total length, 5 mm. in diameter, with the 10 nerves more or less retrorsely hairy, teeth triangular, shortly acuminate, with white membranous margins, 3 mm. long. Petals 11–12 mm. long, white, with a bilobed lamina and well-developed coronal scales.

TANGANYIKA. Masai District : Mt. Ololmoti, Oldonyowass Camp, 16 Sept. 1932, B. D. Burtt 4381 !
DISTR. T2 ; endemic in northern Tanganyika
HAB. Near streams in upland moor, 2490 m.

15. SILENE
L., Sp. Pl. 416 (1753) & Gen. Pl., ed. 5, 193 (1754)

Herbs or subshrubs. Leaves exstipulate, not or only very slightly connate at the base. Inflorescences cymose, panicled, spicate, aggregated-capitulate or reduced even to a solitary flower ; no calycine bracts. Calyx tubular or dilated, 5 teeth, 10 main veins. Petals 5, with a long narrow claw and an entire or usually a bilobed limb. Stamens 5 + 5 in male or herma-phrodite flowers or abnormally reduced in number. Ovary 3- to 5-locular at least in the lower part though usually unilocular in the upper part ; styles 2 to 5 but generally 3, filiform. Capsules dehiscing generally by 3 or 6 teeth or valves.

Annual weeds, with spreading hairs, calyx 0·8–1·0 cm.
  long, entire limb to petals . . . . . 1. *S. gallica*
Perennial herbs, sometimes more or less shrubby at the
  base, glabrous or with short hairs, calyx 1·1–4·4 cm.
  long, bilobed limb to petals :
Flowers in apparently simple rather loose one-sided
  racemes. Calyx 1·1–2·5 cm. (rarely up to 3·5 cm.)
  long . . . . 2. *S. burchellii*
Flowers in a lax divaricate cymose panicle, or an inflorescence reduced from this. Calyx 3·2–4·4 cm.
  long . . . . 3. *S. macrosolen*

1. S. gallica *L.*, Sp. Pl. 417 (1753) ; Mert. & Koch, in Röhl., Deutschl. Fl. 3, 230 (1831) ; Koch, Syn. Fl. Germ. et Helv. 100 (1835). Type : France (LINN, holo. !)

An annual herb with simple or branched stems, 1·4–4·0 dm. tall, with spreading hairs, some very short others up to 2 mm. in length. Leaves oblanceolate or spathulate-oblanceolate, generally rounded and mucronulate at the apex, narrowed to the base, up to 7·7 cm. long and 2·2 cm. broad. Inflorescence with the appearance of a one-sided raceme terminating the stem or its branches, elongating with age, 2–9-flowered. Calyx 8–10 mm. long, enlarging (and especially broadening) in fruit. Petals entire, 9–11 mm. long and with the expanded portion 1·5–3·5 mm. broad, white or pinkish. Capsules on stalks 2–10 mm. long, erect to more or less reflexed. Seeds small, grey to black, minutely tubercled to flat-plated ("armadillo").

UGANDA. Kigezi District : Kachwekano Farm, Feb. 1951, *Purseglove* 3576 !
KENYA. Nakuru District : Njoro, 27 Sept. 1915, *Dowson* 280 !
TANGANYIKA. Lushoto District : Mkusu, 10 Aug. 1936, *Greenway* 4610 !
DISTR. U2 ; K3, 5 ; T3 ; Central and S. Europe, N. Africa, Turkey to Iran, now introduced as a weed into many parts of the world
HAB. Weed of wheat and maize fields, roadsides, waste places, grasslands, 1500–2700 m.

This species has no doubt been introduced by man into East Africa. Of the three varieties usually accepted for the British flora, var. *quinquevulnera* (L.) Koch has not so far been seen from East Africa. Plants that may be referred to var. *gallica* (as *Bogdan* 3183, from Nakuru District, near Molo, 24 July 1951 !) and others that may be con-sidered var. *anglica* (L.) Koch (as *Greenway* 4610) have been received. The varieties, however, are by no means clear-cut and both plants with intermediate development of supposedly diagnostic characters and with various combinations of such characters (branching, inflorescence density, petal size and colouration, length and orientation of fruiting pedicels, etc.) occur both in East Africa and other areas of introduction and in

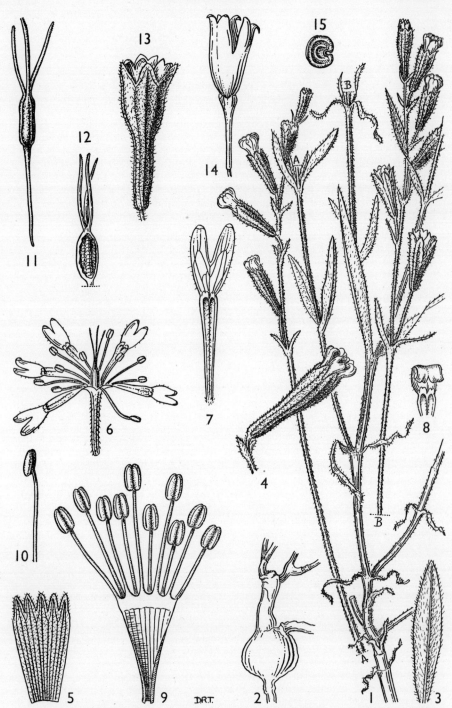

FIG. 13. *SILENE BURCHELLII*—**1,** part of plant, × 1 ; **2,** tuberous root, × ½ ; **3,** leaf, lower surface, × 1 ; **4,** flower, × 2 ; **5,** calyx, opened, × 2 ; **6,** flower with calyx removed to show internode and petals spread out, × 2 ; **7,** petal with limb flattened, × 4 ; **8,** upper part of petal, × 4 ; **9,** stamens and internode cut open (diagrammatic), × 4 ; **10,** stamen, lateral view, × 4 ; **11,** gynoecium and gynophore, × 4 ; **12,** gynoecium with ovary wall removed to show ovules, × 4 ; **13,** persistent calyx enclosing capsule, × 2 ; **14,** dehisced capsule and gynophore, × 2 ; **15,** seed, × 8. **1, 3–14,** from *Haarer* 2291 ; **2,** from *Milne-Redhead* 919 ; **15,** from *Greenway* 6290.

the supposed natural range. The aggregate species would well repay cyto-genetical investigation.

2. **S. burchellii** [*Otth ex*] *DC.*, Prodr. 1 : 374 (1824) ; Fl. Cap. 1 : 128 (1860) ; F.T.A. 1 : 138 (1868) ; Cat. Welw. Afr. Pl. 1 : 49 (1896) ; Burtt Davy in K.B. 1924 : 228 ; Exell & Mendonça in Consp. Fl. Angol. 1 : 113 (1937) ; F.C.B. 2 : 150 (1951). Type : South Africa, Cape of Good Hope, *Burchell* 271 (K, holo. !)

A perennial herb with large more or less ovoid tuberous roots (tubers said to be edible), usually covered with short hairs in all aerial parts ; stems erect or ascending, often branched low down, up to 7 dm. but usually 3–4 dm. tall. Leaves very variable in size and shape, narrowly linear to oblong-lanceolate or oblong-oblanceolate, generally acute at the apex. Flowers apparently in simple, rather loose, one-sided racemes in the uppermost part of the branches with 2–7 flowers ; usually a bare portion of stem between the uppermost leaves and the lowermost flowers. Calyx tubular-clavate, 1·1–2·5 cm. (very rarely up to 3·5 cm.) long (slightly elongating in fruit) ; teeth lanceolate to ovate, obtuse or more or less acute to shortly acuminate. Petals reddish-brown, purple, pink, mauve to white. Capsules on stalks 5–6 mm. long, not reflexed. Seeds flattened, 1·25 mm. diam., not tubercled. Fig. 13.

UGANDA. Ankole District : Kabira, 19 Sept. 1929, *Snowden* 1433 !
KENYA. Naivasha District : summit of Longonot Crater, 17 Dec. 1950, *Verdcourt & Greenway* 403 (narrow leaved variant) !
TANGANYIKA. Mbeya District : Mbozi, 29 Aug. 1933, *Greenway* 3630 (broader leaved variant) !
DISTR. U1–3 ; K1–4, 6 ; T1–8 ; South Africa, Northern and Southern Rhodesia, Angola, Portuguese East Africa, Ruanda-Urundi, southern Belgian Congo, A.-E. Sudan, Ethiopia, Eritrea, Somaliland, Arabia
HAB. Rocky and stony places, screes, upland moor, and grasslands, 1500–4050 m.

SYN. *S. pilosellaefolia* Cham. & Schl. in Linnaea 1 : 41 (1826). Type : South Africa, Cape Province, Knysna, Plettenbergs Bay, *Mund & Maire* (LE, holo. ?)
  S. chirensis A. Rich., Tent. Fl. Abyss. 1 : 44 (1847). Type : Ethiopia, Tigré, Shire, *Quartin Dillon* (P, holo.)
  S. meruensis Engl. in E.J. 48, 382 (1912). Type : Tanganyika, Mt. Meru, *Uhlig* 598 (B, lecto., EA, iso.-lecto. !)

VARIATION. It has not been found possible satisfactorily to subdivide all the large amount of material, here included in one species, into clear cut subspecies or varieties. The type of the species is South African and plants with similar morphology (somewhat suffruticose bases, more or less prostrate or ascending stems, and oblanceolate to spathulate-elliptic leaves) are found in coastal districts of the Cape Province. There is, however, no sharp line of separation between such plants (var. *burchellii*) and others that have been named var. *angustifolia* Sond. and var. *latifolia* Sond.
  In East Africa, plants with very narrow leaves, on the whole, seem to be most frequent in the drier areas, especially of Kenya, but are by no means sharply limited in ecological distribution or geographical range. In addition to variation in leaf width, there is very considerable variation in calyx shape and length and in the length and shape of the calyx lobes. It has not, so far, been possible to obtain any very valid correlations between the variables studied in this complex species (if we so accept it). Most of the specimens seen are in flower and a wide collection of fruits and seeds might throw more light on the nature of the variability though cultural experiments are probably essential before further taxonomic research is worth while.
  A few somewhat extreme variants have been collected. Examples are : Uganda, Karamoja District, Mt. Moroto, 2850 m., *Dale* U. 455 !, with calyces 2·6 to 3 cm. long ; Kenya, Northern Frontier Province, Mt. Endoto, 2400 m., *Jex-Blake in C.M.* 6911, with calyces 3·5 cm. long ; Kenya, Northern Frontier Province, Marsabit, c. 1300 m., *Gillett* 15154 !, with relatively short but broad leaves and linking the var. *gillettii* Turrill (from Ethiopia) to var. *burchellii*.
  With some hesitation a specimen from south-western Tanganyika has been made the type of a new variety under the name *Silene burchellii* var. *macropetala* Turrill in K.B. 1954, 57. This plant is characterized by its tall stems (up to 12 dm.), rather large leaves (3–6·5 cm. long, 0·5–2·1 cm. broad), and petals 1·5 cm. long with the blades

8-9 mm. long and the lobes of the blades 3-4 mm. long.   The type, and only known, specimen came from Songea District (T8), Matengo highlands, above Ugano (presumably Ngano), *Zerny* 518 (W, holo. !).

3. S. macrosolen [*Steud. ex*] *A. Rich.*, Tent. Fl. Abyss. 1 : 44 (1847) ; F.T.A. 1 : 139 ; W.F.K. 14, fig. 16 (1948).   Type : Ethiopia, Semen, near Genausa, *Schimper* 651 (K, iso-lecto. !)

A perennial glabrous herb, often slightly shrubby at base ; stems erect or ascending, up to 8·7 dm. but mostly 3-6 dm. tall, apparently some of the internodes sometimes viscid.   Leaves narrowly linear to oblanceolate-linear, 2-13 cm. long, 1-7 mm. broad, acute to acuminate at the apex, more or less scabrid at the margin towards the base.   Flowers in a lax divaricate sometimes reduced sometimes elongated cymose panicle of 1-8 flowers ; reduced leaves or bracts often up to a short distance below every flower.   Calyx tubular, slightly clavate, 3·2-4·4 cm. long (somewhat elongating in fruit up to 4·7 cm.); teeth semi-ovate, mucronate to acuminate.   Petals white to pink or streaked or stained purple on the outside.

KENYA.   Naivasha District : Mt. Margaret, June 1940, *Bally* 906 !
TANGANYIKA.   Arusha District : Ol Doinyo Sambu, 18 Jan. 1936, *Greenway* 4414 !
DISTR.   K3, 4, 7 ; T2, 3, 5 ; Ethiopia, A.-E. Sudan
HAB.   Grasslands, rocky places, volcanic soils, 1800-3300 m.

SYN.   *S. longitubulosa* Engl., P.O.A. C : 176 (1895).   Type : Tanganyika, Moshi District, Kilimanjaro, north side, *Volkens* 2027 (K, iso. !)

16. DIANTHUS
L., Sp. Pl. 409 (1753) & Gen. Pl., ed. 5, 191 (1754)

Herbs or subshrubs.   Leaves exstipulate, generally " grass-like," more or less connate at base forming a sheath above the often swollen nodes.   Inflorescences terminating the flowering stems, cymose, panicled, or variously aggregated-capitulate ;   a varying number of pairs of calycine bracts surrounding the calyx.   Calyx tubular, 5-toothed, with numerous longitudinal parallel veins.   Petals 5, with a long narrow claw.   Stamens 10 in hermaphrodite flowers or abnormally reduced in number.   Ovary unilocular ; styles 2, distinct from the base.   Capsule cylindrical, oblong, or ovoid, opening by four apical valves or teeth.

D. angolensis [*Hiern ex*] *Williams* in J.B. 24 : 301 (1886) subsp. **orientalis** *Turrill* in K.B. 1954, 49.   Type : Tanganyika, Ufipa District, Chala-Nkunde, *Bullock* 2846 (K, holo. !)

A perennial herb with stiffly erect flowering stems up to 6·2 dm. tall. Leaves narrowly oblanceolate-linear gradually narrowed to the acute or shortly acuminate apex, 2-9 cm. long, 1-9 mm. broad.   Inflorescence elongated, 1·5-2·1 dm. long, with 5-10 flowers ; calycine bracts oblong, oblong-elliptic, elliptic, or obovate-elliptic, at the apex mucronate or shortly acuminate.   Calyx cylindric, 1·8 cm. long.   Petals 2 cm. long, white (rarely, and perhaps only when faded, pale yellow).   Fig. 14.

TANGANYIKA.   Ufipa District : Mwazye, April 1934, *Michelmore* 1039 ! ; Mbeya District : Unyiha, Mkhoma, March 1932, *R. M. Davies* 383 !
DISTR.   T4, 7 ; Belgian Congo, Northern Rhodesia
HAB.   Grassland and in *Brachystegia* woodland, (900-) 1200-1800 m.

Specimens from the Chunya District, E. Rukwa, April 1938, 900 m., *Maclunes* 298 (BM) ! and from the Belgian Congo, Lufuka River, May 1908, *Kassner* 2865 (K and BM) ! verge in some characters towards the subsp. *angolensis*.   More material, especially from localities intermediate between the ranges of the two subspecies is needed to determine whether or not a morphological geographical cline exists.

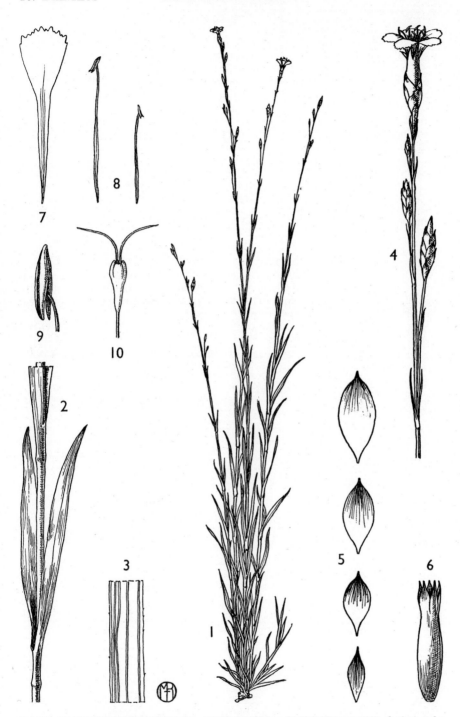

FIG. 14.   *DIANTHUS ANGOLENSIS*, subsp. *ORIENTALIS* from *St. Clair Thompson* 1265—**1**, plant, × ¼ ;
2, part of lower stem, showing leaves, × 1 ; **3**, portion of leaf to show venation, × 2 ; **4**, inflorescence,
× 1 ; **5**, series of bracts, × 2 ; **6**, calyx, × 2 ; **7**, petal, × 2 ; **8**, stamens, × 2 ; **9**, anther, × 10 ; **10**,
gynoecium and gynophore, × 2.

INTRODUCED WEEDS

There is no hard and fast line between casuals and well-established aliens. It is often impossible to be certain what species of the former will in a longer or shorter period of time become the latter. It has seemed most practical to include in this Flora descriptions of introduced species that have now a more or less wide range in Tropical East Africa, as indicated by specimens from three or more distinct areas. There are, in addition, a few specimens from East Africa of species that may or may not become common weeds or ruderals and belong to genera otherwise not represented in the flora. It is desirable that the records of these should be published so that, if they spread, the approximate date of their introduction is known. They are :

Agrostemma githago L. KENYA, Ngong, 13 Dec. 1930, *Hemsted* !

Vaccaria pyramidata Med. KENYA, Eldoret District, near Kapesoret, 25 June 1951, *G. R. Williams* 253 ! ; Olarabel, 48 km. NW. of Thompson's Falls, 1860 m., 1 March 1950, *Lacey* 32 ! var. *grandiflora* (Jaub. & Spach) Čelak.

Melandrium noctiflorum (L.) Fr. UGANDA, Elgon, 2100 m., Sept. 1932, *Mrs. C. Jack* 329 !

Gypsophila elegans M. Bieb. (a large-flowered variant probably the var. *grandiflora* of gardens) : TANGANYIKA, Lushoto District, W. Usambara Mts., Mkuzi, 31 Aug. 1950, *Verdcourt* 341 ! Common in meadow with *Sporobolus, Salvia, Papaver*, etc., 1680 m.

Scleranthus annuus L. KENYA, Aberdare Mts., S. Kinangop, Sept. 1952, *Mrs. R. M. Nattrass* !

INDEX TO CARYOPHYLLACEAE